Advances in Vehicular Networks

Advances in Vehicular Networks

Editors

Barbara M. Masini
Cristiano M. Silva
Ali Balador

MDPI • Basel • Beijing • Wuhan • Barcelona • Belgrade • Manchester • Tokyo • Cluj • Tianjin

Editors
Barbara M. Masini
National Research Council of
Italy and University of Bologna
Italy

Cristiano M. Silva
Universidade Federal
de São João del-Rei
Brazil

Ali Balador
Mälardalen University
Sweden

Editorial Office
MDPI
St. Alban-Anlage 66
4052 Basel, Switzerland

This is a reprint of articles from the Special Issue published online in the open access journal *Journal of Sensor and Actuator Networks* (ISSN 2224-2708) (available at: https://www.mdpi.com/journal/jsan/special_issues/advances_vehicle_networks).

For citation purposes, cite each article independently as indicated on the article page online and as indicated below:

LastName, A.A.; LastName, B.B.; LastName, C.C. Article Title. *Journal Name* **Year**, *Volume Number*, Page Range.

ISBN 978-3-03943-799-3 (Hbk)
ISBN 978-3-03943-800-6 (PDF)

Cover image courtesy of Barbara Mavì Masini.

© 2020 by the authors. Articles in this book are Open Access and distributed under the Creative Commons Attribution (CC BY) license, which allows users to download, copy and build upon published articles, as long as the author and publisher are properly credited, which ensures maximum dissemination and a wider impact of our publications.

The book as a whole is distributed by MDPI under the terms and conditions of the Creative Commons license CC BY-NC-ND.

Contents

About the Editors vii

Barbara M. Masini, Cristiano M. Silva and Ali Balador
Special Issue: Advances in Vehicular Networks
Reprinted from: *J. Sens. Actuator Netw.* **2020**, *9*, 50, doi:10.3390/jsan9040050 1

Afsana Ahamed and Hamid Vakilzadian
Impact of Direction Parameter in Performance of Modified AODV in VANET
Reprinted from: *J. Sens. Actuator Netw.* **2020**, *9*, 40, doi:10.3390/jsan9030040 5

Tassadaq Nawaz, Marco Seminara, Stefano Caputo, Lorenzo Mucchi and Jacopo Catani
Low-Latency VLC System with Fresnel Receiver for I2V ITS Applications
Reprinted from: *J. Sens. Actuator Netw.* **2020**, *9*, 35, doi:10.3390/jsan9030035 21

Dania Marabissi, Lorenzo Mucchi, Stefano Caputo, Francesca Nizzi, Tommaso Pecorella, Romano Fantacci, Tassadaq Nawaz, Marco Seminara and Jacopo Catani
Experimental Measurements of a Joint 5G-VLC Communication for Future Vehicular Networks
Reprinted from: *J. Sens. Actuator Netw.* **2020**, *9*, 32, doi:10.3390/jsan9030032 31

Andreas Xeros, Taqwa Saeed, Marios Lestas, Maria Andreou, Cristiano M. Silva and Andreas Pitsillides
Adaptive Probabilistic Flooding for Information Hovering in VANETs
Reprinted from: *J. Sens. Actuator Netw.* **2020**, *9*, 29, doi:10.3390/jsan9020029 45

Thomas Zehelein, Thomas Hemmert-Pottmann and Markus Lienkamp
Diagnosing Automotive Damper Defects Using Convolutional Neural Networks and Electronic Stability Control Sensor Signals
Reprinted from: *J. Sens. Actuator Netw.* **2020**, *9*, 8, doi:10.3390/jsan9010008 69

Fabio Arena, Giovanni Pau, and Alessandro Severino
V2X Communications Applied to Safety of Pedestrians and Vehicles
Reprinted from: *J. Sens. Actuator Netw.* **2020**, *9*, 3, doi:10.3390/jsan9010003 87

Federico Tonini, Bahare M. Khorsandi, Elisabetta Amato and Carla Raffaelli
Scalable Edge Computing Deployment for Reliable Service Provisioning in Vehicular Networks
Reprinted from: *J. Sens. Actuator Netw.* **2019**, *8*, 51, doi:10.3390/jsan8040051 97

Barbara M. Masini, Cristiano M. Silva and Ali Balador
The Use of Meta-Surfaces in Vehicular Networks
Reprinted from: *J. Sens. Actuator Netw.* **2020**, *9*, 15, doi:10.3390/jsan9010015 111

About the Editors

Barbara M. Masini received a Laurea degree (with honors) and Ph.D. in telecommunications engineering from the University of Bologna in 2001 and 2005, respectively. Since 2005, she has served as a researcher at CNR-IEIIT, and since 2006 she has worked as an Adjunct Professor at the University of Bologna. Her research has been developed within national and international projects in the field of wireless communications, especially focusing on vehicular networks that address issues such as PHY and MAC aspects, as well as real field trials. Masini is also the Editor of IEEE Access and Computer Networks.

Cristiano M. Silva is currently an Associate Professor at the Universidade Federal de São João del-Rei (UFSJ) in Brazil, and the head of the Department of Technology (DTECH). He received his Ph.D. degree in Computer Science from the Universidade Federal de Minas Gerais (UFMG, 2014), Master of Business Administration (IBMEC, 2011), Master's Degree in Computer Science (UFMG, 2005), and Bachelor of Computer Science (UFMG, 2000). His research interests include vehicular networks, Intelligent Transportation Systems, and the application of Artificial Intelligence in urban traffic.

Ali Balador is currently a Senior Researcher and Project Manager at RISE research institute of Sweden and an Assistant Professor at Mälardalen University. He received his Ph.D. degree in Informatics from the Technical University of Valencia in 2016. His research interests include ad hoc and vehicular networks, cyber-physical systems (CPS), Internet of Things (IoT), network protocols, simulation, and security. He was a visiting researcher at Halmstad University, the National Institute of Informatics (NII), and the University of Bologna. He has served as a reviewer for several high-level conferences and journals. He is also involved in the organization of many international conferences.

Editorial

Special Issue: Advances in Vehicular Networks

Barbara M. Masini [1,*], Cristiano M. Silva [2] and Ali Balador [3,4]

1. CNR-IEIIT, v.le Risorgimento, 2, 40136 Bologna, Italy
2. Departamento de Tecnologia, Universidade Federal de São João del-Rei, São João del-Rei 36307-352, Brazil; cristiano@ufsj.edu.br
3. Innovation, Design and Technology (IDT), Mälardalen University, 72123 Västerås, Sweden; ali.balador@mdh.se
4. RISE Research Institute of Sweden, Stora Gatan 36, 72212 Västerås, Sweden
* Correspondence: barbara.masini@ieiit.cnr.it

Received: 10 October 2020; Accepted: 14 October 2020; Published: 17 October 2020

1. Introduction

Connected vehicles are expected to transform the way we travel through the creation of a safe, interoperable wireless communication network among road actors, infrastructures, and objects. Vehicle-to-anything (V2X) communications are, then, fundamental for next-generation active traffic safety and management applications: combined with sensor-based technologies, V2X communications will extend the awareness range of connected and automated vehicles with information received from neighboring vehicles, the infrastructure or vulnerable road users (VRUs).

International consortiums have already agreed on the basic set of applications that need to be implemented to start operating a vehicular network infrastructure at a large scale. Current access technologies, such as IEEE 802.11p (or its European version ITS-G5) and 3GPP LTE-V2X can already succeed in providing most of these so called phase 1 applications, in which vehicles exchange position, speed and other few kinematic information, but they hesitate in the advanced improvement needed to also satisfy the set of requirements foreseen by phase 2 and beyond. Phase 2, in fact, will allow various traffic participants to provide additional information, gained through various on-board sensors like cameras and radar, whereas from phase 3, vehicles will share their intentions so that each vehicle can have a glimpse into the future of other vehicles and, in phase 4, they will exchange and synchronize driving trajectories among each other.

Hence, a fully connected car will require a massive amount of computing power and super-high-speed communication systems, with latency lower than 5 ms and reliability higher than 99.9%.

This Special Issue targeted contributions we deemed relevant to addressing the demand for more reliable and ultra-low-latency vehicular applications. Seven research papers and one review paper have been selected; they range from advancements at the access layer, like the usage of the visible light spectrum to accomodate ultra-low-latency applications, to data dissemination solutions, as well as edge computing, neural-network-based techniques and the use of reconfigurable intelligent surfaces (RIS) to boost throughput and enhance coverage.

Specifically, in [1] the focus was on the improvement in the performance of vehicular networks in terms of delays and packet loss. With this in mind, the authors of [1] proposed a modified version of an ad hoc, on-demand distance vector (AODV) routing protocol to reduce the number of route requests and route reply messages, and thus to reduce the network load and overhead packets. The proposed solution can include additional information in the routing tables, such as the speed, direction and position of the surrounding vehicles. The carry and forward method is applied so that the vehicle nearest to the destination carries the information as long as the source can send the packets to those vehicles going in the same direction, thereby making the route more stable. A two-stage filtering operation is applied to remove the unwanted nodes moving in the opposite direction. Simulation results show a 1.4% reduction in packet loss, an 11% reduction in the end-to-end delay, and an increase in throughput.

A couple of works are dedicated to future scenarios in which traditional radio communications are not able to support the stricter demands on the underlying V2X connectivity technology, in terms of latency, reliability, bandwidth and capacity. Frequently sending information about all on-board sensor observations may saturate the capacity of traditional V2X communication technologies, especially in dense urban environments with a large number of equipped vehicles, as expected, as the penetration ratio of cooperative and automated vehicles (CAVs) increases. Visible light communication (VLC) is then addressed as a complementary technology to support the future demands of data-hungry cooperative sensing applications.

Paper [2] presents a characterization of a low-cost, low-latency VLC prototype for infrastructure-to-vehicle (I2V) communication. The system consists of a legacy traffic light as a transmitter and a photodetector as a receiver. The latter is equipped with low-cost Fresnel lenses as condensers to increase the optical gain in the system at the receiver. The system is capable of active decode and the relaying of information to further incoming units. The experimental characterization of amplitude and Packet Error Rate (PER) for the proposed system demonstrated the feasibility of an error-free I2V communication (corresponding to a PER lower than 10^{-5}) whithin 50 m. Furthermore, the prototype can be used for both broadcast and beaconing transmission modes. This low-cost VLC-based system could offer sub-millisecond latency in the full active decode and relay process for distances of few tens of meters, which makes it suitable for integration in Cellular-V2X (C-V2X) and 5G platforms.

In [3], VLC is jointly used with 5G to provide real-time information to vehicles. This paper presents a field-test that integrates 5G communication capabilities with VLCs to test the end-to-end latency that can be offered by this integrated communication network in a vehicular scenario.

In [4], the concept of information hovering for data dissemination over a mobile set of peers is addressed to make the given information available to all vehicles within a confined geographical area in a specific time period. The paper proposes a strategy based on epidemic routing in the hovering area, and probabilistic flooding outside it to avoid some vehicles never receiving the content dedicated to them due to eventual partitions and disconnected areas created by low traffic density. To this aim, vehicles outside the hovering area serve as bridges towards partitions, leading to high reachability. The work highlights the adaptive feature of the protocol, where the rebroadcast probability in partitions is adaptively regulated based on estimates of the density of vehicles in the hovering area. The performance of the proposed scheme is evaluated in realistic conditions, using a section of the road network in cities of Washington as the reference model in all simulations.

The important topic of vulnerable road users (VRUs) is faced in [5], where vehicle-to-pedestrian (V2P) communications are considered to support road safety by enabling all road users to exchange information. This paper presents an architectural design to support V2P communications and to provide interfaces for vehicle and pedestrian in safety-oriented applications; in addition, Reference [5] proposes a couple of implementable applications for pedestrian safety that could be implemented in tablets and smartphones. The proposed solution is based on the integration of Wi-Fi and IEEE 802.11p in the on-board unit (OBU) of vehicles so that the vehicle can communicate via Wi-Fi with the pedestrian, sending reliable and efficient warnings, whereas IEEE 802.11p is used for V2V communications. Interesting discussions on actual hardware and software solutions are reported, indicating the potentialities and limits of actual wireless technologies in vehicles' and users' smartphones and discussing potential solutions that could be made available when the pedestrian device will be equipped with the most suitable wireless access technologies for vehicular communication scenarios, such as C-V2X and 5G.

Since the connected vehicles can provide a large variety of services, they could be based on different wireless access technologies and they need different network requirements, such as low latency, high computational capacity, and high reliability. In [6], the authors proposes an integer linear program (ILP) to solve the joint deployment problem of baseband processing and edge computing with reliability against single-node failure in a cloud radio access network (C-RAN) to minimize the nodes in which processing capabilities must be installed while ensuring that latency and optical link (i.e., maximum wavelengths over fibers) constraints are not violated. To overcome the computational complexity of classical optimization approaches, a hybrid based on both heuristic and ILP deployment strategy is also proposed. The algorithm performs a first phase, in which the initial set of nodes as candidates for host baseband and edge computing functions is reduced and a suboptimal solution is provided. Then, a second phase is executed for optimization purposes. The latter approach is shown to provide results close the optimal ones while considerably reducing computational time.

The work in [7], instead, investigates the suitability of convolutional neural networks (CNN) for the diagnosis of defective automotive dampers. In fact, chassis system components such as dampers have a significant impact on vehicle stability, driving safety, and driving comfort. Currently, in addition to the driver's perception, there is only periodic human inspection for monitoring the vehicle's chassis system state. However, this is error-prone, expensive and implies periods of unmonitored driving between inspections and cannot be applied in future self-driving cars. Therefore, an automated monitoring and diagnosing system is of high importance. Authors' experiments were conducted with different parameters regarding the size of the receptive field, the size of the pooling layer and the network depth of the CNN architecture, and the resulting kernel weights of the trained networks were analyzed. To ensure a broad applicability of the generated diagnosis system, only signals of a classic electronic stability control (ESC) system, such as wheel speeds, longitudinal and lateral vehicle acceleration, and yaw rate were used. A structured analysis of data pre-processing and CNN configuration parameters were investigated in terms of the defect detection result. The results show that simple fast Fourier transformation (FFT) pre-processing and configuration parameters resulting in small networks are sufficient for a high defect detection rate.

Finally, in the review paper [8], the unprecedent importance of a smart environment besides connected and automated vehicles is addressed. A smart environment could drastically enhance the performance of wireless access technologies and connected vehicles. Hence, as more and more connected and autonomous vehicles hit the road, we need to smarten up the streets on which we travel, for example, by coating the environment with reconfigurable meta-surfaces, able to customize the radio wave propagation and opportunistically reach the target. Meta-surfaces are thin electromagnetic meta-materials, typically deeply sub-wavelength in thickness and electrically large in transverse size. They are composed of sub-wavelength scattering particles that can revise the Snell's law redirecting the radio waves in the desired direction and can achieve this run time, changing the redirection of the waves over time. Hence, the major difference between a surface and a meta-surface is the capability of the latter of shaping the radio waves according to the generalized Snell's laws: for example, the angles of incidence and reflection of the radio waves are not necessarily the same. Beyond the meta-surface, what is really challenging and stimulating is the use of a reconfigurable meta-surface, where the scattering particles are not fixed from the manufacturing phase, but can be modified depending on the stimuli that the meta-surface receives from the external world. The work in [8] describes the main characteristics of meta-surfaces and highlights the potential uses of reconfigurable meta-surfaces, especially when adopted in vehicular environments, focusing specifically on cooperative driving and vulnerable road user (VRU) detection. An analytical model (validated by simulation) is also presented to demonstrate the improvement that a reconfigurable meta-surface can provide in reducing the collision probability when random access to the medium is adopted for vehicle-to-vehicle (V2V) communications.

It is hoped the readers will appreciate this Special Issue enhancing their knowledge in the area of enabling technologies for connected and automated vehicles and helping them advance their on-going research and innovation activities.

Funding: This research received no external funding.

Conflicts of Interest: The authors declare no conflict of interest.

References

1. Ahamed, A.; Vakilzadian, H. Impact of Direction Parameter in Performance of Modified AODV in VANET. *J. Sens. Actuator Netw.* **2020**, *9*, 40. [CrossRef]
2. Nawaz, T.; Seminara, M.; Caputo, S.; Mucchi, L.; Catani, J. Low-Latency VLC System with Fresnel Receiver for I2V ITS Applications. *J. Sens. Actuator Netw.* **2020**, *9*, 35. [CrossRef]
3. Marabissi, D.; Mucchi, L.; Caputo, S.; Nizzi, F.; Pecorella, T.; Fantacci, R.; Nawaz, T.; Seminara, M.; Catani, J. Experimental Measurements of a Joint 5G-VLC Communication for Future Vehicular Networks. *J. Sens. Actuator Netw.* **2020**, *9*, 32. [CrossRef]
4. Xeros, A.; Saeed, T.; Lestas, M.; Andreou, M.; Silva, C.M.; Pitsillides, A. Adaptive Probabilistic Flooding for Information Hovering in VANETs. *J. Sens. Actuator Netw.* **2020**, *9*, 29. [CrossRef]
5. Arena, F.; Pau, G.; Severino, A. V2X Communications Applied to Safety of Pedestrians and Vehicles. *J. Sens. Actuator Netw.* **2020**, *9*, 3. [CrossRef]
6. Tonini, F.; Khorsandi, B.M.; Amato, E.; Raffaelli, C. Scalable Edge Computing Deployment for Reliable Service Provisioning in Vehicular Networks. *J. Sens. Actuator Netw.* **2020**, *8*, 51. [CrossRef]
7. Zehelein, T.; Hemmert-Pottmann, T.; Lienkamp, M. Diagnosing Automotive Damper Defects Using Convolutional Neural Networks and Electronic Stability Control Sensor Signals. *J. Sens. Actuator Netw.* **2020**, *9*, 8. [CrossRef]
8. Masini, B.M.; Silva, C.M.; Balador, A. The Use of Meta-Surfaces in Vehicular Networks. *J. Sens. Actuator Netw.* **2020**, *9*, 15. [CrossRef]

© 2020 by the authors. Licensee MDPI, Basel, Switzerland. This article is an open access article distributed under the terms and conditions of the Creative Commons Attribution (CC BY) license (http://creativecommons.org/licenses/by/4.0/).

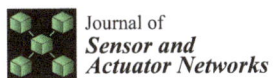

Article

Impact of Direction Parameter in Performance of Modified AODV in VANET

Afsana Ahamed [1] and Hamid Vakilzadian [2,*]

[1] Department of Electrical Engineering, Arkansas Tech University, Russellville, AR 72801, USA; aahamed@atu.edu
[2] Department of Electrical and Computer Engineering, University of Nebraska-Lincoln, Lincoln, NE 68588, USA
* Correspondence: hvakilzadian@unl.edu

Received: 13 June 2020; Accepted: 24 August 2020; Published: 3 September 2020

Abstract: A vehicular ad hoc network (VANET) is a technology in which moving cars are used as routers (nodes) to establish a reliable mobile communication network among the vehicles. Some of the drawbacks of the routing protocol, Ad hoc On-Demand Distance Vector (AODV), associated with VANETs are the end-to-end delay and packet loss. We modified the AODV routing protocols to reduce the number of route request (RREQ) and route reply (RREP) messages by adding direction parameters and two-step filtering. The two-step filtering process reduces the number of RREQ and RREP packets, reduces the packet overhead, and helps to select the stable route. In this study, we show the impact of the direction parameter in reducing the end-to-end delay and the packet loss in AODV. The simulation results show a 1.4% reduction in packet loss, an 11% reduction in the end-to-end delay, and an increase in throughput.

Keywords: AODV; end-to-end delay; packet loss ratio; throughput; VANET

1. Introduction

Intelligent traffic systems (ITS) and vehicular ad hoc networks (VANETs) are established to permit communication among vehicles to decrease traffic congestion and increase safety. A VANET uses moving cars as wireless routers (nodes) to establish a mobile network for communication [1]. The network is created by applying the principles of mobile ad hoc networks (MANET) to build a wireless network for exchanging data spontaneously. In this technology, only the vehicles equipped with wireless transceivers can exchange data with neighboring vehicles to transfer data packets to destinations that are not within direct communication range.

Compared to a MANET, a VANET has higher, structured mobility and a broad coverage area. It requires little or no power and has no service fee. However, it needs continuous data exchange to update the road structure, such as the number of lanes, the number of cars on the road, the number of roadside units (RSUs), etc. In addition, a VANET requires fast, reliable environmental data for proper and safe vehicle navigation and speed control. Because of the continuously changing number of nodes and their mobility, the throughput is low in a VANET; and packet loss is high due to connection failure. The topology created in a VANET is dynamic and not uniformly distributed. Therefore, maintaining the quality of service (QoS) is crucial; and routing of the packets is a major challenge, especially when bandwidth is limited [2].

Among the routing protocols suggested for VANETs, the Ad hoc On-Demand Vector (AODV) [3] offers rapid adaptation to changes in dynamic link, has low overhead for processing and memory usage, and provides low network utilization to determine unicast routes to a destination within the network.

The AODV is an on-demand or reactive routing protocol for a wireless ad hoc network. It initiates a route discovery when it needs to transmit data packets to a destination node, and it does not have any path towards the destination node prior to transmitting the data packets. This network consists of three procedures: (1) route discovery process, (2) route message generation, and (3) route maintenance. Since the route is only created when needed, it requires less overhead as compared to proactive routing protocols. Therefore, one of the main advantages of an AODV is the low overhead required for the data packets. In an AODV, the routing information is not updated after a specific period, which is also bandwidth efficient. This approach decreases the effects of inactive routes along with the need for route maintenance for unused routes. To improve the performance of an AODV in a VANET, Abedi et al. [4] proposed the use of direction information for each node as a parameter for selecting the next hop during a route discovery phase. However, their study did not include end-to-end (E2E) delay, throughput (TH), and packet delivery ratio (PDR). The broadcasting effect to mitigate flooding problems was also not addressed.

Wang et al. [5] used fuzzy logic and fuzzy control to make a routing decision. They developed fuzzy control-based AODV (FCAR) routing protocols. By comparing AODV and FCAR, they concluded that FCAR outperforms AODV in E2E delay and packet drop percentage. One of the main caveats of this study is that the packet drop rate of FCAR increases significantly when the network size is more than 70 nodes or the speed of the nodes (cars) is more than 10 m per second. Therefore, further study is required to reduce the packet drop rate for performance improvement.

Ding et al. [6] suggested improved AODV routing protocols to enhance stability and decrease the control overhead of the route. In this study, they used two-step optimization in the route discovery and route selection. One of the main contributions of this method is that it reduces the number of broken links. However, this method does not provide satisfactory performance in terms of packet delivery ratio. In addition, it does not consider other performance matrices, such as throughput and E2E delay, when considering the performance of the proposed method.

Sun et al. [7] proposed a global positioning system (GPS)-based AODV (GBAODV) routing protocol to establish a route. They constrained the flooding of AODV routing packets using GPS devices to improve routing performance. They found that their GBAODV method reduces the network load more than the AODV method. As a result, the number of broken links and the packet loss ratio are reduced. The average E2E delay is also shorter when using GBAODV than AODV. However, when they considered different highway scenarios, the performance of the GBAODV method for packet loss was not satisfactory (more than 10%). Their study was also based on a small number of nodes (only eight nodes).

Yu et al. [8] incorporated vehicles' movement information into the route discovery process based on an AODV for VANET applications to improve the reliability of the routing protocols (more stable route). They considered the total weight of the route (based on the position metrics) (TWR) and the estimated expiration time in their proposed protocol to achieve more stable routing. They found that their proposed protocol reduces the routing load even more and ensures more stable connections. Nevertheless, their proposed protocol does not show significant improvement in the percentage of packet drops.

He et al. [9] proposed a new vehicular reliability model that uses vehicle movement information and channel state information to improve the reliability of routing in VANETs. They extended the AODV routing protocol by proposing a reliable ad hoc, on-demand distance vector routing protocol AODV-L. They compared the traditional AODV routing with AODV-L using Objective Modular Network NESTbed in C++ (OMNeT++). They found that the AODV-L routing protocol outperformed the AODV in terms of PDR and E2E delay. Nevertheless, this study only focused on rural highways.

Feyzi et al. [10] suggested fuzzy logic to improve the efficiency of routing protocols in AODV. They used direction, speed, and distance of vehicles to the destination as inputs to a fuzzy logic controller. Their proposed protocol outperformed the original AODV for average E2E and both AODV and FCAR for the packet delivery rate. Nevertheless, the proposed protocol underperformed FCAR regarding the average E2E delay.

Wang et al. [11] proposed an efficient message routing framework to optimize the message delivery throughput from vehicles to RSUs. They developed a mathematical model to examine the asymptotic throughput scaling of VANETs. However, the throughput of their method was poor (0.16 packets/sec). To improve the QoS of the vehicular network under different road scenarios (urban, rural, and highway), optimization of performance matrices (i.e., throughput, the percentage of packet loss, E2E delay, and the network coverage) is essential.

To increase reliability and resource usage efficiency based on medium access control (MAC) protocol, Bazzi et al. [12,13], suggested applying orthogonal frequency division multiple access (OFDMA) for alert message flooding in VANETs. They compared their protocols with other benchmark MAC protocols. In [14], Karabulut et al. proposed an OFDMA-based efficient cooperative MAC (OEC-MAC) protocol for VANETS. The OEC-MAC protocol ensured an increase in throughput with a delay of 100 ms for safety messages. The network connection reliability was raised by reducing the packet dropped rate.

In our study, we modified the AODV routing protocols to include the information on speed, direction, and position of vehicles. The routing tables now include the additional direction information of the nearby vehicles. In [4], Abedi et al. used the direction parameter to select the next-hop only. In our method, the carry and forward methods are applied so that the nearest vehicle to the destination carries the information as long as the source can send the packets to those vehicles going in the same direction, thereby making the route more stable. Filtering is done in two stages to remove the unwanted nodes moving in the opposite direction.

In the above-mentioned studies, the packets were sent to the closest vehicles in all directions; but in our approach, we used a priority-based routing where the packets were routed to the nodes that were closest and were moving in the same direction. This new method uses two filtering steps and, therefore, increases the throughput to more than 10%, compared to other state-of-the-art techniques, reduced the packet loss to 2%, and improved the E2E delays as compared to other state-of-the-art methods.

2. AODV and Modified AODV Routing Algorithms

The AODV [15] routing protocol is reactive, where routes are determined only when needed. Ad hoc On-Demand Vector is bandwidth efficient because it works only on demand. Figure 1 presents the AODV routing protocol message exchanges.

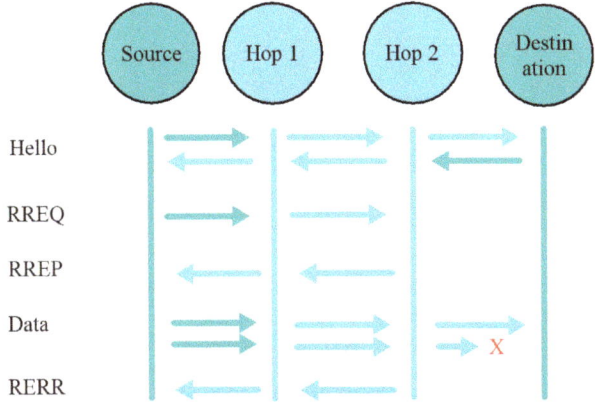

Figure 1. AODV messaging.

In this protocol, each mobile node identifies other neighborhood nodes by flooding or by receiving a local broadcast known as a Hello message. At this time, to provide an immediate response to the requester for establishing new routes, the routing tables of the neighboring nodes are modified with response time of local movements. The main objectives of this routing protocol are [15]:

1. To broadcast discovery packets (Hello) and receive acknowledgment when necessary.
2. To distinguish between general topology maintenance (route management) and local connectivity management (detection of neighborhood).
3. To inform the neighboring mobile nodes about the changes in local connection.

Hello messages [15] are utilized to detect and to observe the nearby neighbors. When Hello messages are used, every active node in the network periodically transmits a Hello message to its neighbors. These messages permit the detection of a connection or link break. The protocol uses different types of messages, shown below, to discover and maintain the links (broken links or selecting a different path). They are:

1. Route Request (RREQ)
2. Route Reply (RREP)
3. Route Reply Acknowledgement (RREP-ACK)
4. Route Error (RERR)

A source broadcasts [15] an RREQ to transfer data to an unknown destination. Therefore, a route from a node to the source is created at the intermediate node after the reception of an RREQ. If the receiving node does not receive the RREQ, the source rebroadcasts it. When the destination node is the receiving node or has a recent route from the source to the destination, it creates an RREP. As the RREP circulates, intermediate nodes generate routes to the destination. If the source receives the RREP, it registers the route to the destination in its routing table. Then it starts directing data. The source chooses the path with the shortest hop count after receiving multiple path information.

When the data is transferred from the source to the destination, nodes update their timers, which is related to the time the route is used between the source and destination to maintain the routing table. When a route is inactive (for about 2000 ms), the node invalidates the route and removes it from its routing table.

The RREP-ACK message [15] is generally sent in response to an RREP message to complete the route discovery cycle. If a link failure is detected while transferring the data, the node sends an RERR message to the source in a hop-by-hop approach. As the RERR is headed for the source, the intermediate nodes invalidate routes to inaccessible destinations. The source also invalidates the route after receiving the RERR and initiates the route discovery again.

An AODV uses a destination sequence number (DSN) [15] to avoid counting to infinity to be loop free (no duplication of packet sending). The destination includes the destination sequence number and any route information when sending messages to requesting nodes. A node broadcasts an RREQ to all network nodes to discover a route as long as it finds the destination or another node with the most recent route to the destination. The requesting node then selects the best route based on the DSN (an AODV uses a DSN to determine the most recent path to a destination). The entries in the routing table are related to sequence numbers generated by the destination. A DSN acts as a route timestamp, confirming a fresh route. When an intermediate node receives an RREQ packet, it compares its DSN with that in the RREQ packet. If the stored DSN is greater than the current DSN in the RREQ packet, the existing route is considered to be up to date. At that point, an RREP is sent back to the source, and the route (discovered) is made available.

An AODV has three main phases of operation [15]:

1. *Route discovery* is initiated when a source node needs to communicate with another neighbor node. Each node keeps two counters: broadcast_id and node_sequence_number and (so that individual nodes can be recognized and the duplication of packets can be avoided) an RREQ. When the source node issues an RREQ, the broadcast_id is incremented. Each neighbor satisfies

the RREQ by either sending back an RREP or rebroadcasting the RREQ after increasing the hop count in its neighborhood.
2. *Route maintenance* does not have any impact on the route to the destination. It reinitiates the route discovery process when the source node moves. When the destination nodes or intermediate nodes move, an RREP is sent to the affected nodes. Periodic Hello messages or link-layer acknowledgments (LLACKS) (far less latency) can be used to determine node movements. If a node is not reachable, the special RREP is sent back toward the source, containing a new sequence number and hop count.
3. *Route message generation* use a local connectivity to maintain a list of active nodes within its neighbors in one of two ways. First, if a node obtains a broadcast message from a nearby node, it updates its routing table, containing connectivity information. Second, when a neighbor does not send any packets within the Hello interval, it broadcasts a Hello message that includes its identity and its DSN.

An AODV has a routing table management system where information about neighbor nodes is updated periodically. Information about short-lived routes is also stored in the routing table, such as the routes created to store reverse paths towards nodes originating RREQs temporarily. The route entry table in the AODV [15] includes the following information on its neighbor:

- Destination sequence number
- Destination IP address
- Valid destination flag
- Next hop
- Hop count
- State and routing flags (e.g., repairable, being repaired, valid, and invalid)
- List of precursors
- Network interface
- Lifetime (i.e., deletion or expiration time of the route)

In AODV, there is no special security to prevent attacks at high latency because the route discovery is reactive and requires more time. This results in high packet loss. Since it does not allow the handling of unidirectional links, a single route request packet may result in multiple route reply packets, leading to huge routing overhead. Periodic beaconing (to determine the active/alive state of a node) leads to unnecessary bandwidth consumption, which leads to low throughput. The high mobility of the vehicles may also cause more link breakage, which increases the delay in transferring the data packets.

To maintain QoS and to overcome a flooding problem, low overhead is required when replying to a single route request. Moreover, the periodic beaconing time can be optimized depending on the size and number of nodes (cars) of the network. The route discovery phase can be modified to work more effectively with less time delay. The AODV routing protocols can be modified to give priority to a node to handle unidirectional links. Due to the highly dynamic nature of the network, the routing table can be modified and filtered to give priority to nodes based on the location information of each node.

Modified AODV

In this study, we proposed an improved AODV routing protocol for VANETs by introducing two optimization steps: (1) a route discovery phase and (2) a route selection phase. The current location, speed, and direction information about vehicles is included for performance optimization while keeping the E2E delays to a minimum. We assumed a Manhattan model in our research experiment environment. In this model, a vehicle can go straight with 50% probability and left or right, with a 25% probability each. We also assumed a highway model where the vehicles can move in the opposite direction. The AODV packets carry more information on each vehicle to keep the other vehicles more up to date about each node on the road. We assumed automated cars with automated cruise control and wireless sensors will be used to prevent accidents. Therefore, the RREP and RREQ packets now have three pieces of information: velocity, direction, and position of each source node. The routing table in the vehicles is updated with this information so that all of the nearby vehicles can be known to each other.

This method will certainly not increase the size of a packet. However, the direction information will provide more filtering options (to make the route more stable) for the vehicles. In our model, we have assumed two different situations:

1. *Nonaccident mode:* In this case, each vehicle will update its routing table for nearby cars; but it will transmit the information to those cars which are moving in the same direction as well as all nearby RSUs within 92 m.
2. *Accident mode:* In this case, each vehicle will update the information in its routing table so that if there is an accident, the information will be sent in all directions and to nearby RSUs so that all of the vehicles can update their routing tables accordingly. Here the filtering process will be divided in two different ways. In the first half, the information will be sent to all of the vehicles within 92 m. If an accident is on the opposite side of the road, the vehicle will keep moving and updating the information for nearby vehicles. However, if it is on the same side of the road, the vehicle will stop (if it is immediate to the point of the accident and there is no free lane to move into), or it will reduce its speed and change into an available lane for safety and broadcast information.

The RREQ packet format is changed in a modified AODV so that the reserved eight bits can be used for assigning the direction in which the source is moving in terms of the degree of rotation with reference to the equator. Therefore, each node of the network has the direction information along with the location of the destination. In the case of an accident, the information can be filtered and prioritized. This information will be given to the cars moving in the same direction and close to the accident. Therefore, the accident information will be passed to the nearby nodes with more ease.

The RREQ message format is presented in Figure 2, and the fields include [15]:

Type	1 for RREQ, 2 for RREP
J	A flag which is reserved for joining
R	A flag which is reserved for repair
G	A gratuitous RREP flag when a gratuitous RREP has to be unicast to the node specific in destination IP address field
D	A destination flag when the destination is responding to the RREQ by itself
U	A flag for unknown DSN
Reserved	Source moving direction
Hop Count	The total number of hops from the source IP address to the destination IP address during the request

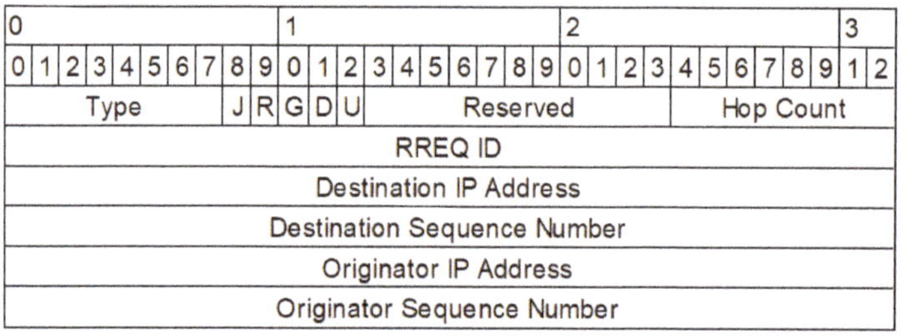

0										1										2										3	
0	1	2	3	4	5	6	7	8	9	0	1	2	3	4	5	6	7	8	9	0	1	2	3	4	5	6	7	8	9	1	2
Type										J	R	G	D	U	Reserved									Hop Count							
RREQ ID																															
Destination IP Address																															
Destination Sequence Number																															
Originator IP Address																															
Originator Sequence Number																															

Figure 2. RREQ message format.

The nearest node is calculated using the Pythagoras formula. If the location of node A is (λ_1, φ_1) and the location of node B is (λ_2, φ_2), then the shortest distance between them is calculated as [16]

$$d = 2R \sin^{-1} \sqrt{\sin^2 \frac{(\varphi_1 - \varphi_2)}{2} + \cos \varphi_1 \cos \varphi_2 \sin^2 \frac{(\lambda_1 - \lambda_2)}{2}} \quad (1)$$

where R is the earth radius, φ is the latitude, and λ is the longitude.

If the latitudes of nodes A and B are φ_1 and φ_2, respectively, the direction between them is calculated as

$$\theta = \varphi_1 - \varphi_2 \quad (2)$$

In addition, the modified AODV routing table contains the direction information column for each nearby node that will be used to filter the nodes to pass data with little or no delay. Priority will be given to those nodes which have a minimum angle θ. Here, we assume that if the angle is less than 20 degrees, then the nodes are moving in the same direction [4].

3. Simulation Environment

We assessed the performance of the proposed routing protocols in OMNeT++ and used MATLAB® to analyze the simulation data. Recently, the combination of simulation of urban mobility (SUMO) and OMNeT++ has been used to build the network communication in a VANET. We combined the network with a traffic simulator named SUMO, which is used to reproduce the network protocols in the VANET. The mobility of each node in this simulation environment is controlled by SUMO, and the position is forwarded periodically to the network simulator.

In a VANET, the outcomes of a simulation are not entirely reliable and mostly guided by the mobility model in use. In this context, Sommer et al. [17] suggested the necessity for a robust simulation environment for VANETs. This led to the advancement of Veins (vehicles in network simulation) [18]. Veins is an open-source simulation environment that uses two simulators: (1) SUMO for conducting microscopic road traffic simulation and (2) OMNeT++ for conducting the network simulation, as shown in Figure 3.

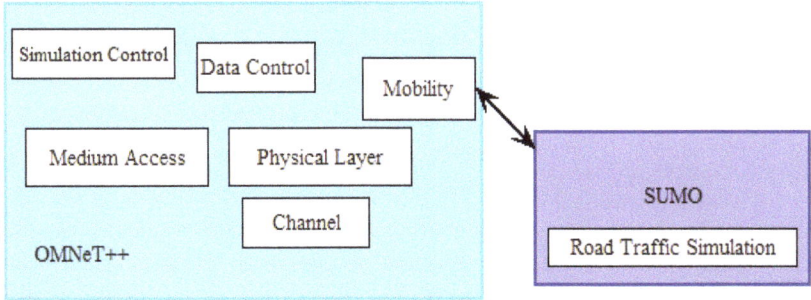

Figure 3. Objective Modular Network NESTbed in C++ (OMNeT++) with simulation of urban mobility (SUMO) [16].

The road map for VANET simulation is imported from openstreetmap.org [19] by manually choosing the required region or area and exporting to the .osm option to download a map.osm file. The resultant file is an XML file with an OpenStreetMap-defined region. We used Java Open Street Map Editor (JOSM) to edit the map.osm file manually. Apart from the road network, the imported and edited map.osm also includes other information, such as rivers, walkways, power distribution lines, bicycle routes, parks, and other features that are outside the scope of this study. A brief description of the simulation tools used in this research is given below:

3.1. SUMO

Simulation of Urban Mobility is an open-source traffic simulation software. When a vehicle or node moves through a given road network, SUMO can handle a given traffic demand and permits addressing a large set of traffic scenarios. In SUMO, each vehicle is modeled explicitly, having its own route, and moves autonomously through the system. There are also several alternatives to introduce randomness to the simulation [20].

3.2. JOSM

Java Open Street Map (OSM) Editor, a desktop application, was developed by Immanuel Scholz. It is currently maintained by Dirk Stöcker. It is an open street map for a JAVA platform that supports loading standalone GPS tracks from an OSM database to load and edit existing nodes. The imported map.osm also includes other information (e.g., rivers, walkways, power distribution lines, railway networks, and bicycle routes, etc.) apart from the road network. In this study, JOSM was used to edit the map. We delete unnecessary nodes; and, if required, we add new nodes with new routes and infrastructures to the simulation environment. Therefore, the final map.osm file contains only the road network and infrastructure that are necessary for the simulation environment [21].

3.3. OMNeT++

Objective Modular Network NESTbed in C++ is a discrete event simulator. It is used for modeling wired and wireless communication networks as well as microprocessors, multiprocessors, and other distributed or parallel computing systems. It is based on the C++ language, and it consists of different basic modules that communicate with each other by passing messages among them. These basic modules are used to create larger modules [22].

The Network Description (NED) language is used to simulate models in OMNeT++. The component modules described in this language can be assembled to create a compound (system) module. Models written in NED can be simulated by one of the network simulators, such as MiXiM [23], INET [24]/INETMANET [25], or Veins [26]. Veins is the open-source vehicular network simulator. The execution of a model by Veins is controlled by OMNeT++ while interacting simultaneously with a road traffic simulator (SUMO). Various submodules in Veins take care of setting up, running, and monitoring the simulation.

When performing intravehicular communication (IVC) evaluations, both traffic and network simulators run in parallel since they are connected via a Transmission Control Protocol (TCP) socket. This communication network protocol is known as Traffic Control Interface (TraCI) [27].

The vehicle movement in SUMO is directed by the movement of nodes in an OMNeT++ simulation environment. We have also used Plexe [28] for the movement of an automated vehicle in OMNeT++. Plexe is an extension of Veins in OMNeT++. It allows realistic and accurate simulation of the automated car. It structures real-time vehicle dynamics and several cruise control models, enabling the implementation of large-scale and mixed scenario control systems along with different networking protocols.

4. VANET Component Model

The following are some of the essential models of VANET components that were used in the simulation.

4.1. Two-Ray Interference Model

The Two-Ray Ground Reflection Model is a radio propagation model used to predict the path losses between transmitting and receiving antennas of the nodes. The received signal has two components: the line-of-sight (LOS) component and the multipath component, which is formed by a single ground-reflected wave. This model considers the impact of both the direct path and the ground reflection during information propagation. The path loss is included to model the propagation of information in the vehicular network. The received power is

$$P_r \propto \frac{1}{d^4} \quad (3)$$

where d is the distance between the transmitting and receiving antennas.

For any nonlinear (not smooth) unobstructed stretch of road, depending on the distance, the transmission faces constructive or destructive interference for its ground reflection. The two-ray ground model only considers that path loss increase for distances over approximately 900 m. Therefore, to capture ground reflection, VANET is comprised of a Two-Ray Interference model [29].

4.2. Obstacle Shadowing Model

The signal shadowing effects are heavily impacted by radio transmission. Therefore, obtaining the real signal is important in VANET in both urban and suburban areas, where buildings and infrastructure block radio propagation. In the simulation, we include a calibrated and validated obstacle shadowing model against realistic obstacle measurement [18]. This model accurately captures the effect of blocking transmission by large buildings. In other words, while strong transmissions can be hindered by the presence of infrastructure in the line of sight, weak transmissions can be blocked by something as small as a short wall.

4.3. Adaptive Cruise Control (ACC) Model

An ACC system offers driver comfort and ease by providing cruise control in a high mobility traffic environment. Adaptive cruise control systems improve highway safety. About 94% of highway accidents occur through human error [30], while a small percentage of highway accidents are caused by equipment failure or weather conditions (such as slippery roads). Since an ACC system possibly decreases driver liability and changes driver function with an automated process, it is estimated that the implementation of an ACC system will reduce the number of crashes and hazards [28] on the road.

An ACC system with two modes of steady-state operation (i.e., speed control and vehicle spacing control) is shown in Figure 4.

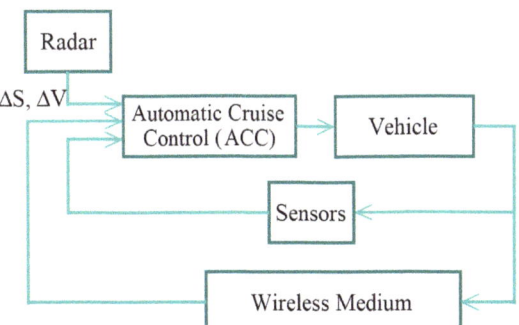

Figure 4. Adaptive cruise control (ACC) with vehicle [28].

4.4. Roadside Unit (RSU)

Roadside units are essential nodes to provide communication support between vehicles. This support may be in the form of reporting an accident, safety warnings, and traffic congestion. A model of an RSU is shown in Figure 5. Each RSU is equipped with a network interference card (NIC) and the application layer software, appl. The NIC unit also includes the AODV router.

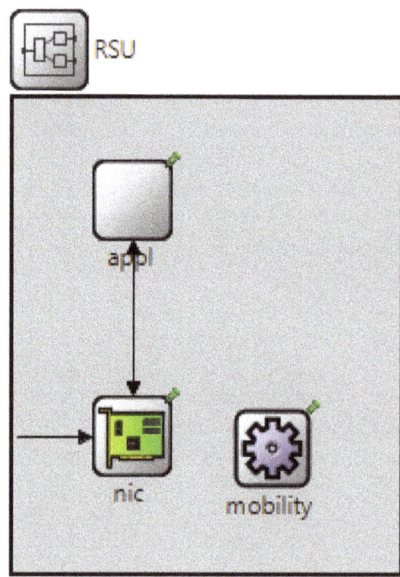

Figure 5. Roadside unit (RSU) model.

5. Results and Discussion

To compare our results with published articles, we used the map of Erlangen in SUMO to generate the traffic information. This map is used primarily in VANET simulations in most of the state-of-the-art methods. The mobility model was generated using OMNeT++, and then the simulations were run by varying the number of vehicles. The simulation was run 20 times to have a statistically significant output.

Figure 6 shows the roads and the existing infrastructure of the area. The black lines are the roads, and the red-colored areas indicate the infrastructure.

Figure 6. View of Erlangen map in SUMO.

The simulation result in OMNeT++ is shown in Figure 7, where the blue dots are the moving nodes in the same area in SUMO. A summary of the simulation parameters is provided in Table 1.

The percentage of packet loss versus the number of cars is presented in Figure 8. It can be seen from the plot that the packet loss increased with an increase in the number of cars.

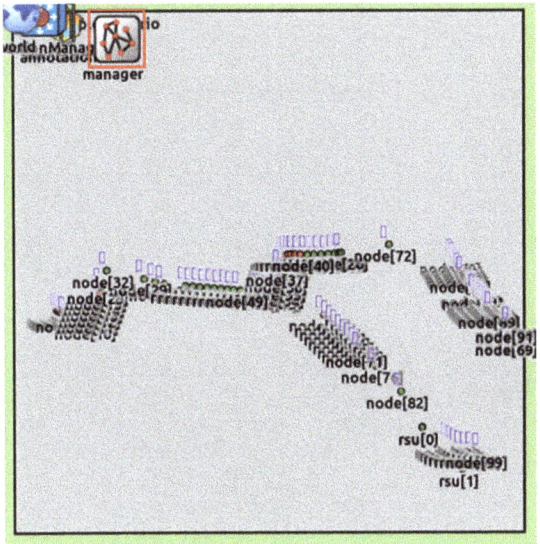

Figure 7. Simulation in OMNeT++.

Table 1. Summary of simulation parameters.

Parameters	Value
Network simulator	OMNeT++
Simulation time	1000 s
Number of nodes	100–800
Simulation coverage volume	2500 m (L) × 2500 m (W) × 50 m (H)
MAC layer	802.11 p
Propagation model	Two ray interferences
Mobility model	Random trip and Intelligent car
Number of runs	20
Velocity	20–40 mps

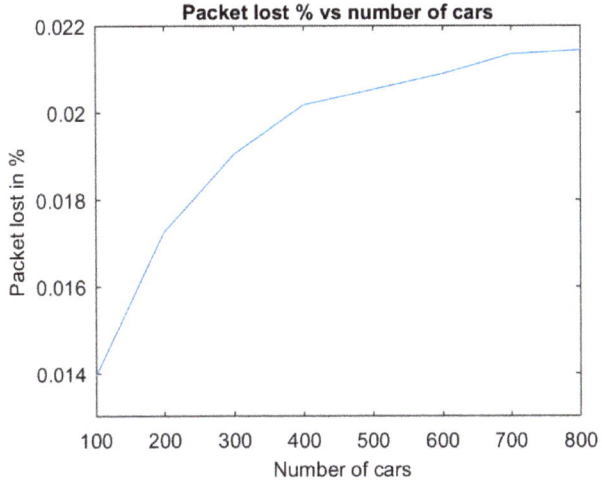

Figure 8. Percentage of packet loss vs. the number of cars.

The throughput vs. the number of cars is shown in Figure 9. It can be seen from this graph that the throughput increased with the number of cars due to the high network activities with more cars.

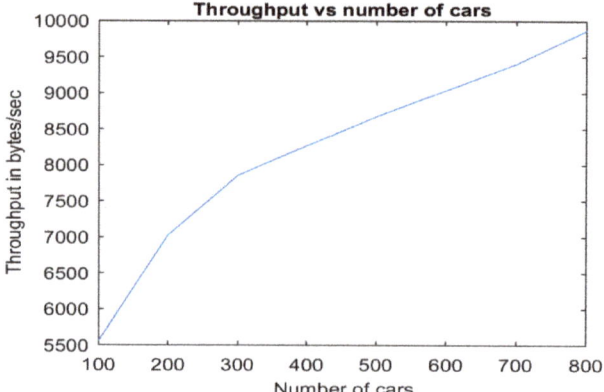

Figure 9. Throughput vs. the number of cars.

We examined the performance regarding packet loss and packet delivery ratios during the simulation; and the results are provided in Table 2 with 100 nodes and in Table 3 with 200 nodes, along with the results of other methods [4–6,9]. In both cases, the inclusion of the direction parameter outperformed the others by about 1.3% under the same environmental conditions.

Table 2. Percentage of packet loss with 100 nodes.

Methodology	Packet Loss	Packet Delivery Ratio
Our method	1.4%	98.6%
Wang et al. [5]	2.75%	97.25%
Sun et al. [7]	10%	90%

Table 3. Percentage of packet loss with 200 nodes.

Methodology	Packet Loss	Packet Delivery Ratio
Our method	1.7%	98.3%
Wang et al. [5]	3%	97%
He et al. [9]	3%	97%
Ding et al. [6]	7%	93%

The comparison of an average E2E delay is presented in Table 4. As shown, our method displayed a decrease of 11% in delay compared with the other methods.

Table 4. E2E delay performance.

Methodology	Average E2E Delay
Our method	178 ms
Wang et al. [5]	200 ms
Sun et al. [7]	400 ms
He et al. [9]	300 ms
Feyzi et al. [10]	225 ms
Wang et al. [11]	250 ms

The improved performance is the result of using two filtering steps in route discovery and route selection procedures. In this case, each node has the most updated routing information; therefore, it also has fewer broken links. The packet drop rate and end-to-end delay also decreased due to a lower number of broken links in a stable route. These processes also reduced the number of RREQ and RREP messages, which increased throughput.

6. Conclusions and Future Work

In this research paper, we simulated the performance analysis of VANET using OMNeT++ and SUMO by collecting the map from OpenStreetMap. We modified the AODV routing protocols to reduce the number of RREQ and RREP messages by adding direction parameters and two-step filtering. The two-step filtering process reduced the number of RREQ and RREP packets, reduced the packet overhead, and helped selection of the stable route. We also compared the performance metrics in terms of E2E delay, packet delivery ratio, and throughput with other state-of-the-art techniques.

The inclusion of direction parameters along with two-step filtering in route discovery and route selection procedures reduced the average end-to-end delays and packet loss by 11% and 1.3%, respectively, as compared to other state-of-the-art methods. We also found an improvement in throughput that was due to the combined effect of direction parameters and filtering in the same direction to find the stable route, which reduced the packet loss during communication. Therefore, the proposed method increases the QoS in VANET, providing more safety on the road.

In our future research, we will study and propose new protocols that might help to improve the performance of protocols such as dynamic source routing with a grid or geographic location service protocol for VANETs concerning road safety in real-time applications.

Author Contributions: Conceptualization, methodology, software, and preparation of original draft by A.A.; supervision, review, recommendation for changes, and editing by H.V. All authors have read and agreed to the published version of the manuscript.

Funding: The APC is funded by the Department of Electrical and Computer Engineering, University of Nebraska-Lincoln.

Conflicts of Interest: The authors declare no conflict of interest.

Nomenclature

ACC	Automatic cruise control
AODV-L	AODV link expire time
AODV	Ad hoc On-demand Distance Vector
DSN	Destination sequence number
E2E	End-to-end
FCAR	Fuzzy control-based AODV routing
GBAODV	Global positioning system (GPS)-based AODV
GPS	Global positioning system
ITS	Intelligent traffic system
IVC	Intervehicle communication
JOSM	Java OpenStreetMap
LLAKS	Link-layer acknowledgments
LOS	Line-of-sight
MAC	Medium access control
MANET	Mobile ad hoc network
NED	Network description
NIC	Network interface card
OFDMA	Orthogonal frequency division multiple access
OMNeT++	Objective Modular Network NESTbed in C++
OSM	OpenStreetMap
PDR	Packet delivery ratio

PDR	Packet dropped rate
QoS	Quality of service
RSU	Roadside unit
RERR	Route error
RREP	Route reply
RREP-ACK	Route reply acknowledgment
RREQ	Route request
SUMO	Simulation of Urban Mobility
TCP	Transmission Control Protocol
TH	Throughput
TraCI	Traffic Control Interface
TWR	Total weight of the route
VANET	Vehicular ad hoc network
Veins	Vehicles in network simulation

References

1. Benslimane, A. Localization in Vehicular Ad Hoc networks. In *2005 Systems Communications (ICW'05, ICHSN'05, ICMCS'05, SENET'05)*; Institute of Electrical and Electronics Engineers (IEEE): Piscataway, NJ, USA, 2005.
2. Ahamed, A.; Vakilzadian, H. Issues and challenges in VANET routing protocols. In Proceedings of the 2018 IEEE International Conference on Electro/Information Technology (EIT), Rochester, MI, USA, 3–5 May 2018; pp. 723–728.
3. Perkins, C.E.; Royer, E.M. Ad-hoc on-demand distance vector routing. In Proceedings of the WMCSA'99. Second IEEE Workshop on Mobile Computing Systems and Applications, New Orleans, LA, USA, 25–26 February 1999.
4. Abedi, O.; Fathy, M.; Taghiloo, J. Enhancing AODV routing protocol using mobility parameters in VANET. In Proceedings of the 2008 IEEE/ACS International Conference on Computer Systems and Applications, Doha, Qatar, 31 March–4 April 2008; pp. 229–235.
5. Wang, X.-B.; Yang, Y.-L.; An, J.-W. Multi-metric routing decisions in VANET. In Proceedings of the 2009 Eighth IEEE International Conference on Dependable, Autonomic and Secure Computing, Chengdu, China, 12–14 December 2009; pp. 551–556.
6. Ding, B.; Chen, Z.; Wang, Y.; Yu, H. An improved AODV routing protocol for VANETs. In Proceedings of the 2011 International Conference on Wireless Communications and Signal Processing (WCSP), Nanjing, China, 9–11 November 2011; pp. 1–5.
7. Sun, Y.; Chen, Y.; Xu, Y. A GPS enhanced routing protocol for vehicular Ad-hoc network. In Proceedings of the 2011 IEEE International Conference on Robotics and Biomimetics, Phuket, Thailand, 7–11 December 2011; pp. 2096–2101.
8. Yu, X.; Guo, H.; Wong, W.-C. A reliable routing protocol for VANET communications. In Proceedings of the 7th International Wireless Communications and Mobile Computing Conference, Istanbul, Turkey, 4–8 July 2011; pp. 1748–1753.
9. He, Y.; Xu, W.; Lin, X. A stable routing protocol for highway mobility over vehicular ad-hoc networks. In Proceedings of the 2015 IEEE 81st Vehicular Technology Conference (VTC Spring), Glasgow, UK, 11–14 May 2015; pp. 1–5.
10. Feyzi, A.; Sattari-Naeini, V. Application of fuzzy logic for selecting the route in AODV routing protocol for vehicular ad hoc networks. In Proceedings of the 23rd Iranian Conference on Electrical Engineering, Tehran, Iran, 10–14 May 2015; pp. 684–687.
11. Wang, M.; Shan, H.; Luan, T.H.; Lu, N.; Zhang, R.; Shen, X.; Bai, F. Asymptotic throughput capacity analysis of VANETs exploiting mobility diversity. *IEEE Trans. Veh. Technol.* **2014**, *64*, 1. [CrossRef]
12. Bazzi, A.; Masini, B.M.; Zabini, F. On the exploitation of OFDMA properties for an efficient alert message flooding in VANETs. In Proceedings of the 2013 IEEE International Conference on Communications, Xi'an, China, 12–14 August 2013; pp. 5094–5098.
13. Bazzi, A.; Zanella, A.; Masini, B.M. An OFDMA-based MAC protocol for next-generation VANETs. *IEEE Trans. Veh. Technol.* **2014**, *64*, 4088–4100. [CrossRef]

14. Karabulut, M.A.; Shah, A.F.M.S.; Ilhan, H. OEC-MAC: A novel OFDMA based efficient cooperative MAC protocol for VANETS. *IEEE Access* **2020**, *8*, 94665–94677. [CrossRef]
15. Perkins, C.; Belding-Royer, E.; Das, S. *Ad hoc On-Demand Distance Vector (AODV) Routing*; RFC Editor: Santa Barbara, CA, USA, 2003.
16. Mahmoud, H.; Akkari, N. Shortest path calculation: A comparative study for location-based recommender system. In Proceedings of the 2016 World Symposium on Computer Applications & Research (WSCAR), Cairo, Egypt, 12–14 March 2016; pp. 1–5. [CrossRef]
17. Sommer, C.; German, R.; Dressler, F. Bidirectionally coupled network and road traffic simulation for improved IVC analysis. *IEEE Trans. Mob. Comput.* **2010**, *10*, 3–15. [CrossRef]
18. Sommer, C.; Dietrich, I.; Dressler, F. Realistic simulation of network protocols in VANET scenarios. In Proceedings of the 2007 Mobile Networking for Vehicular Environments, Anchorage, AK, USA, 11 May 2007; pp. 139–143. [CrossRef]
19. OpenStreetMap Wiki Contributors, "OpenStreetMap", OpenStreetMap Wiki. 2004. Available online: http://www.openstreetmap.org (accessed on 7 August 2020).
20. Krajzewicz, D.; Erdmann, J.; Behrisch, M.; Bieker, L. Recent development and applications of SUMO—Simulation of urban mobility. *Int. J. Adv. Syst. Meas.* **2012**, *5*, 128–138.
21. OpenStreetMap Wiki Contributors, JOSM, OpenStreetMap Wiki. 2017. Available online: http://wiki.openstreetmap.org/w/index.php?title=JOSM&oldid=1492146 (accessed on 7 August 2020).
22. Varga, A.; Hornig, R. An overview of the omnet++ simulation environment. In Proceedings of the 1st International ICST Conference on Simulation Tools and Techniques for Communications, Networks and Systems, Marseille, France, 3–7 March 2008.
23. OMNeTpp/mixim, GitHub. 2016. Available online: https://github.com/OMNeTpp/mixim (accessed on 7 August 2020).
24. INET Framework, OMNeT++. 2012. Available online: http://inet.OMNeTpp.org/ (accessed on 7 August 2020).
25. aarizaq/inetmanet-3.x, GitHub. 2017. Available online: https://github.com/aarizaq/inetmanet-3.x (accessed on 7 August 2020).
26. Sommer, C. Veins. 2006. Available online: http://veins.car2x.org (accessed on 7 August 2020).
27. Wegener, A.; Piórkowski, M.; Raya, M.; Hellbrück, H.; Fischer, S.; Hubaux, J.-P. TraCI: An interface for coupling road traffic and network simulators. In Proceedings of the 11th Communications and Networking Simulation Symposium, New York, NY, USA, 14–17 April 2008.
28. Segata, M.; Joerer, S.; Bloessl, B.; Sommer, C.; Dressler, F.; Cigno, R.L. Plexe: A platooning extension for veins. In Proceedings of the 2014 IEEE Vehicular Networking Conference (VNC), Padeborn, Germany, 3–5 December 2014; pp. 53–60.
29. Henderson, T. Two-Ray Ground Reflection Model, isi.edu. 5 November 2011. Available online: https://www.isi.edu/nsnam/ns/doc/node218.html (accessed on 7 August 2020).
30. Singh, S. *Critical Reasons for Crashes Investigated in the National Motor Vehicle Crash Causation Survey*; National Highway Traffic Safety Administration: Washington, DC, USA, 2015.

© 2020 by the authors. Licensee MDPI, Basel, Switzerland. This article is an open access article distributed under the terms and conditions of the Creative Commons Attribution (CC BY) license (http://creativecommons.org/licenses/by/4.0/).

Article

Low-Latency VLC System with Fresnel Receiver for I2V ITS Applications

Tassadaq Nawaz [1,2], Marco Seminara [3], Stefano Caputo [4] and Lorenzo Mucchi [4] and Jacopo Catani [2,3,*]

1. Dept. of Physics and Astronomy, University of Florence, 50121 Florence, Italy; tassadaq.nawaz@unifi.it
2. National Institute of Optics-CNR (CNR-INO), 50125 Florence, Italy
3. European Laboratory for NonLinear Spectroscopy (LENS), University of Florence, 50121 Florence, Italy; seminara@lens.unifi.it
4. Dept. of Information Engineering, University of Florence, 50121 Florence, Italy; stefano.caputo@unifi.it (S.C.); lorenzo.mucchi@unifi.it (L.M.)
* Correspondence: jacopo.catani@ino.cnr.it

Received: 26 June 2020; Accepted: 17 July 2020; Published: 22 July 2020

Abstract: This work presents a characterization of a low-cost, low-latency Visible Light Communication (VLC) prototype for infrastructure-to-vehicle (I2V) communication for future Intelligent Transportation Systems (ITS). The system consists of a regular traffic light as a transmitter (the red light is modulated with the information), and a photodetector as a receiver. The latter is equipped with low-cost Fresnel lenses as condensers, namely, 1″ Fresnel and 2″ Fresnel, to increase the optical gain of the system at the receiver. The system is capable of Active Decode and Relay (ADR) of information to further incoming units. The experimental characterization of amplitude and Packet Error Rate (PER) for the proposed system has been performed for distances up to 50 m. The results show that by incorporating the 2″ Fresnel lens in the photodetector, an error free (PER $\leq 10^{-5}$) I2V communication is established up to 50 m. Furthermore, the prototype can be used for both broadcast and beaconing transmission modes. This low-cost VLC-based system could offer sub-millisecond latency in the full ADR process for distances up to 36 m, which makes it suitable for integration in Cellular-V2X (C-V2X) and 5G platforms.

Keywords: infrastructure-to-vehicle; vehicle-to-vehicle; Intelligent Transportation Systems; Visible Light Communication; Fresnel lenses

1. Introduction

The capabilities of existing Intelligent Transportation Systems (ITS) can significantly be improved by enabling advanced low-latency infrastructure to vehicular (I2V), vehicular to vehicular (V2V) and vehicular to infrastructure (V2I) communications, which could offer active road safety applications by assisting drivers in critical moments. In order to enable fast and robust vehicular communications several technologies and techniques [1] have been proposed and tested, but most efforts focus on dedicated short range communications (DSRC) and IEEE standard 802.11p, which forms the regulations for wireless access in vehicular environments (WAVE) [2,3]. These standards use dedicated frequency bands for ITS in Europe and the United States to provide the potential solutions for future implementations of communication-based ITS safety applications [4,5]. On the other hand, the Third Generation Partnership Project (3GPP) introduced a new standard commonly referred to as LTE-V, LTE-V2X, or cellular V2X. This vehicular communications standard is based on well known cellular standard long-term evolution (LTE), which supports a side link or V2V communications using LTE's direct interface named PC5 [6–8]. For safety-critical applications, vehicles must be equipped with ultra-reliable and ultra-low latency communication systems to share information with infrastructures and other vehicles for triggering appropriate actions, for example, through their electronic driving assistance systems.

In this context, LED-based visible light communication (VLC) has recently drawn huge attention from the communications community due to exciting features which are lacking in common RF-based communication systems, such as, the high degree of integrability in existing infrastructures and the intrinsic directionality of the VLC channel. The latter could allow for the implementation of agile and highly-reconfigurable ad-hoc I2V and V2V subnetworks via directional interconnectivity among endpoints [9], and becomes a prominent factor in recent laser-based VLC implementations, besides the very large attainable bandwidths exceeding the GHz range [10,11]. VLC uses the unregulated visible light spectrum (400–790 THz) for communication, which can be exploited for either high data rates for indoor Li-Fi or pervasive broadcast of short information packets with very low latency, which is especially important in ITS safety-critical applications [12,13]. Furthermore, VLC can find applications where radio waves do not provide a suitable solution for wireless connectivity, for instance, due to the presence of strong electromagnetic (EM) interference (heavy industrial facilities), or of particularly sensitive EM equipment (planes, operating rooms, etc.).

Focusing on VLC in ITS, two types of receivers, namely, camera-based and photodiode-based are considered in the literature. Camera based prototypes exploit the tendency of modern vehicles to be equipped with cameras for lane and pedestrian detection, and the limited framerate and noise performances could. However, to overwhelm such issues, several works introduced complex image processing architectures with dedicated image processing hardware [14] and tracking mechanisms [15] to attain long cast and reasonable bitrate, often using non-regulated, custom LED arrays as the source to increase the bitrate of the VLC link [16–18]. High-speed cameras can also be used to improve the performance of camera-based VLC systems, but they are too expensive and computationally complex to be exploited in the current automotive industry. On the other hand, low-cost dedicated photodetectors are quite efficient regarding noise performances and data rates, and they seem to provide a good, lower-cost alternative for vehicular communications. Recent works focused on either V2V applications with data rates ranging from 20 kbaud [19] up to 115 kbaud [20], or I2V [21–23] implementations. In particular, in ref. [24], an I2V2V prototype featured message relaying capabilities to further nodes, with a total distance limited to 18 m and a maximum rate of 15 kbaud. None of the above-mentioned works employed a regular traffic light infrastructure as a VLC source. Furthermore, the end-to-end latency of the VLC chain has not been measured. In photodiode-based VLC, the optical gain of the receiver is typically provided by an additional optical collection system, that reduces the FOV of the receiver, hence limiting the interfering light coming from sides. This system englobes optical concentrators or lenses to focus the incoming light on the photodetector, hence, more energy is collected on the detection area. Consequently, more Signal-to-Noise ratio (SNR) is available at the receiver to perform a demodulation of bits. Therefore, optical lenses can be used to increase the communication distance as well as the performance, in terms of Packet Error Rate (PER), of VLC systems. Fresnel lenses offer good optical performances at a competitive price with respect to standard molded glass aspheric lenses, featuring reduced thickness and weight. Their usage in receiver stages has been reported in former works, reporting attainable distances of few meters at 1kbps in a laboratory prototype [25], whilst other works focused on design methods for Fresnel lenses in indoor VLC applications [26]. However, a detailed characterization of performances of photodiode-based VLC systems is lacking, and it would be essential towards the deployment of VLC in realistic ITS applications.

In this paper, we perform an experimental evaluation of a low-cost, Fresnel-lens based VLC I2V prototype system (see Figure 1) for ITS employing a commercial, regulatory LED traffic light as the source. Data are sent to an incoming receiving node, which consists of a photodetector, incorporating two different low-cost, most commonly available plastic lenses, namely, 2″ Fresnel (FR2) and 1″ Fresnel (FR1). The receiving node is able to perform Active Decode and Relay (ADR) of messages received from the traffic light to a further incoming node. Data are transmitted at two rates, i.e., 115 and 230 kbaud. Our results show that $FR2$ outperforms the $FR1$ lens while offering an error free communication

(PER = 10^{-5}) for a range of distances of 6–50 m for 115 kbaud and 6–36 m for 230 kbaud. We find critical configurations in the relative position of traffic light and receiving units, possibly helping in the design of novel, realistic ITS based on VLC wireless links. We also report sub-millisecond latencies in the whole ADR process for both beaconing and message broadcasting in ITS protocols, making it virtually suitable for integrability with 5G technology.

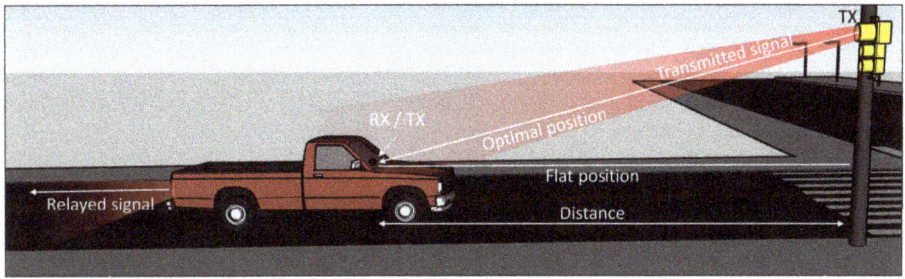

Figure 1. Proposed prototype for infrastructure-to-vehicle (I2V) communication: Traffic light transmits messages to the vehicle, which decodes and compares the message with a stored one and relays only if whole message is received correctly. Two configurations, Flat and Optimal are considered for performance analysis of the prototype.

The rest of the paper is organized as follows: Section 2 introduces the experimental setup. The experimental results are presented and discussed in Section 3 and then we conclude this paper in Section 4.

2. Experimental Configuration

The experiments are carried out in a 55 m-long corridor in the Department of Physics and Astronomy of the University of Florence. The experimental setup used to validate our newly developed prototype is shown in Figure 2. A block diagram of the VLC system is reported in Figure 2d. The equipment is composed of two main units, Transmitter unit (TX) and Receiver unit (RX).

TX is composed of two sub-units, modulator and light source. The modulator/encoder is realized through a low-cost open source microcontroller-based platform (Arduino DUE) and analog custom-designed current modulator. A commercial traffic light, provided by the company ILES srl in Prato (https://ilessrl.com), is used as an optical source, which converts the electrical signals to optical signals. The transmitter directly casts the UART data stream through On-Off Keying (OOK) modulation and Non-Return-to-Zero (NRZ) data coding [27].

RX unit is composed of a photodetector mounted on 105 cm-high tripod, emulating car dashboard height, a comparator/digitizer and a demodulator. The photodetector consists of a photodiode (Thorlabs PDA36A2) with a variable gain and of a lens used to concentrate the light on the area of the photodiode to increase the SNR at the receiver. Two lenses, namely, $FR1$ (Thorlabs FRP125) and $FR2$ (Thorlabs FRP232) with different size and focal length, are used as condensing elements. A digital comparator with hysteresis converts the received analog signal to digital, then feeding it to Arduino DUE, which performs demodulation and recovers the information. A 1 Gs/s digital oscilloscope is used for signal recording and analysis. Details on the electro-optical TX-RX system are given in ref. [13], as they are out of the scope of this paper.

The TX board produces a predefined bit sequence through one of its UART ports, and arranges it into a packet of 6 bytes (2 bytes preamble + 4 bytes payload). Each packet is preceded by 3 equalization bytes, to limit the effects of low-frequency transients on the digitizer/comparator stage, which would otherwise lose the first bits of each packet due to the presence of the AC coupling stage (see below and also Figures 5 and 6). Data packets are encoded using NRZ line coding, carrying 1 bit/symbol and allowing a maximum data rate of 230 Kbps. These encoded data are then fed to the current modulator, which modulates the current supplying the red LED lamp (Lux Potentia, 3 LEDs, 6.5 W nominal) of the traffic light using OOK modulation.

The transmitted signal propagates through the optical channel towards the RX stage. The latter is AC-coupled before the first amplification stage to reject the DC stray light components such as sunlight (during the day a large amount of sunlight can enter the corridor through the windows) or low-frequency, 100 Hz intensity fluctuations coming from artificial lights which are present in the corridor. The amplified signal is then digitized, decoded and analyzed by the Arduino DUE board, which decodes the incoming message and compares it with a stored reference message to determine the PER. As the RX is equipped with ADR capabilities [13], if a message is correctly received, it can be forwarded to a second modulator for relaying towards the incoming unit through, for example, rear lamps (not taken into account in the present work).

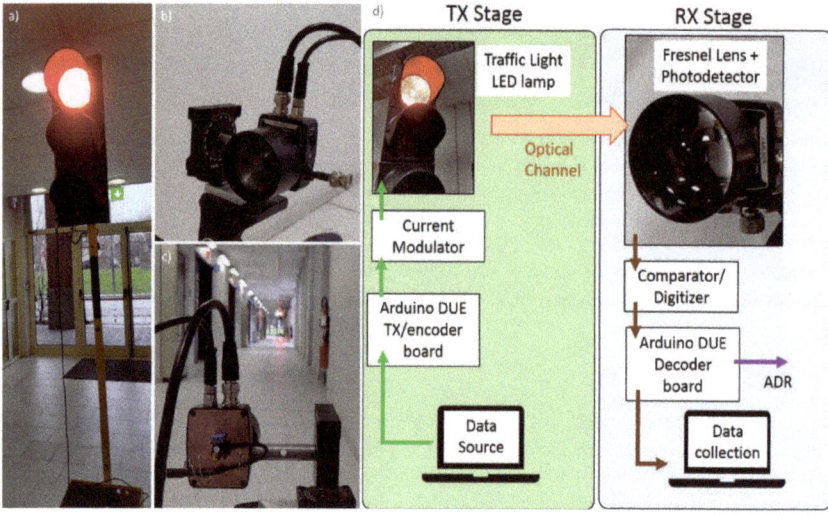

Figure 2. Experimental setup: (**a**) Transmitter unit: a standard traffic light that is used as a transmitter. (**b**) Receiver unit: photodetector, which consists of the Fresnel lens and an AC-coupled photodiode with a variable gain. (**c**) Receiver is positioned in front of traffic light and two system configurations, Optimal and Flat are used for performance analysis of the prototype. (**d**) Block diagram of the complete Visible Light Communication (VLC) chain.

3. Results and Discussion

Experiments are performed by placing the RX stage in various positions in front of a traffic light, for two system configurations; (a) the photodetector optical axis always aims to the traffic light red lamp, known as Optimal system configuration; (b) the photodetector axis is parallel to the floor, known as Flat system configuration (See Figure 1). Whilst the first configuration features the best SNR (as no angular misalignment between TX and RX optical axes is involved), the latter configuration is the most similar to a realistic scenario where no adaptive tracking mechanism is involved. Two low-cost Fresnel plastic lenses $FR1$ with a focal length of 25 mm and $FR2$ with focal length 32 mm are used in RX to focus the incoming light from traffic lamp on the area of the photodiode. The RX is mounted on a precision vertical rotational platform. The performances of the system are evaluated through PER. The reason why we adopt the PER characterization rather than the Bit Error Rate (BER) in this specific application relies on the fact that in ITS applications one of the most critical aspects is the statistical value of latency in critical messages delivery. The latter, in turn, can be directly connected to the average number of lost packets [13], which is what is measured in a PER analysis.

Also, the PER analysis is typically less demanding from the hardware point of view and does not need for a post-processing of data. Anyhow, we notice that for low error rates and uniform error distribution, PER provides a very good approximation for BER through PER = NBER, where N is the packet length [28]. We measure PER as a function of communication distances between TX and RX, and we quantify the minimum latency of the whole ADR process, which is here defined as the time elapsed between the transmission of first bit in a packet sent by TX and the last bit of the corresponding relayed packet by the ADR stage of RX. The system is tested for both $FR1$ and $FR2$ lenses in each configuration in order to compare their performances. The traffic light is placed in a static position and photodetector is moved from 1 to 50 m. To retrieve the PER, a predefined message is sent for 10^5 times from traffic light to photodetector. In such configuration, an error-free communication gives an upper bound for PER to 10^{-5}.

In the first set of experiments, the received signal amplitude is recorded for various communication distances up to 50 m for both $FR1$ and $FR2$, and results are shown in Figure 3. It can be noticed from Figure 3a that in Optimal system configuration, when no angular misalignment is present, the received signal strength with $FR2$ is globally higher than with $FR1$. This is due to the reason that the diameter of $FR2$ is twice the diameter of $FR1$. The larger optical gain allows the condenser stage to collect more light and convey a larger intensity on the area of photodiode. After a steady increase in the region 3–6 m, which is related to the intensity map cast by the traffic light lamp [29], the signal strength reduces as distance increases for both lenses reaching 12 mV and 40 mV at 50 m for $FR1$ and $FR2$, respectively. On the other hand, in Flat system configuration, as shown in Figure 3b, for near distances up to 25 m, the amplitude of the received signal with $FR1$ is significantly higher than the one received with $FR2$. This due to the fact that lenses with shorter focal length provide the RX stage with larger Field of View (FOV), hence more easily collecting light coming from off-axis sources, as it happens, for example, when a car is very close the traffic light. However, for long distances the trend is similar to that of Optimal system configuration (Figure 3a), the $FR2$ outperforms the $FR1$. At large distances, indeed, the relative angle between RX and TX optical axis reduces to very low values, de facto making this configuration to approach the Optimal case, where no angular misalignment is involved. An amplitude of 10 and 25 mV is recorded at 50 m for $FR1$ and $FR2$, respectively. It is worth noticing that Figure 3 highlights a remarkable deviation from the pristine $1/d^2$ decay behavior in the optical intensity. The $1/d^2$ dependency is related to the solid angle given by the input aperture of the RX stage optical element when dealing with spherical sources. Here, the presence of beam-shaping elements (lenses) at the optical source, along with the variable TX-RX and emission angles on the measurement grid make the $1/d^2$ decay not necessarily a valid approximation in general (see also [29]), especially at short distances where the angles in play are large. We also note that one could eventually obtain an indication of the expected performances of the system using the green lamp as TX (not available at the time of measurements) by simply rescaling the amplitude measured for the red lamp by a correction factor (0.6) which could be retrieved from the photosensitivity VS wavelength graphs on the constructor's website (www.thorlabs.com). Partial compensation for this decrease in the received amplitude is provided by the larger nominal green lamp power (9 W), to be compared to the 6.5 W provided by the red lamp. Neglecting the different height among the red and green lamps slots in the traffic light case, we expect a global correction factor in the received amplitude of $\simeq 0.83$ when a green lamp is used.

We characterize the PER in a second set of experiments. As PER depends on the received signal strength, it only depends, given a certain baudrate, on the detected signal amplitude, which we measure after the RX stage. Hence, we first measure PER as a function of received signal's amplitude for two baud rates, 115 and 230 kbaud, respectively. Then, this calibration procedure allows us to retrieve the experimental characterization of PER by accurately measuring the signal amplitude map along the whole measurement grid (6–50 m). Figure 4 reports such analysis as a function of distance, for both $FR1$ and $FR2$ lenses, in the Optimal and Flat configurations (panels a–b and c–d, respectively), and for both 115 kbaud (a–c panels) and 230 kbaud (b–d panels). Error bars correspond to a variation of

1 mV in the detected amplitude at RX. We verified in key points at 36, 42 and 50 m that the calibration of PER obtained via the amplitude method agrees with the value obtained through direct measurement of PER within error bars. For better comparison with future outdoor implementations, we notice that our indoor setting could be prone to possible effects of multipath reflections. In our low-baudrate regime, the phase delay introduced by such processes would be negligible with respect to the minimum bit time, hence making eventual reflections positively contribute to the received amplitude, possibly leading to slightly better PERs with respect to the outdoor configuration. However, we expect that eventual reflections would give an appreciable contribution only in the low-SNR regime, i.e., in the very near and very far-field regions.

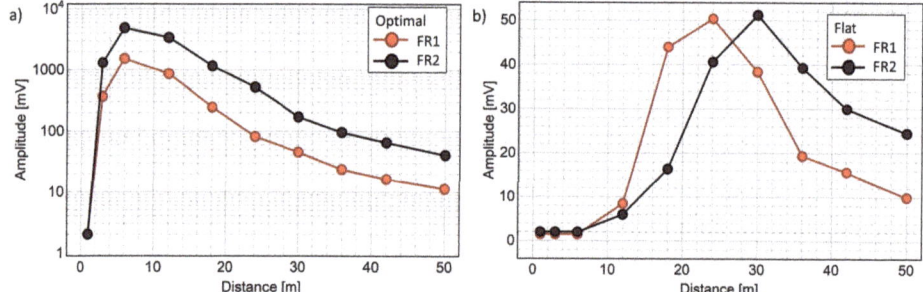

Figure 3. Amplitude of received signal as a function of distance for two optical lenses ($FR1$ and $FR2$) and two receiver positions, (**a**) shows the Optimal configuration in which receiver is always pointing towards the transmitter, and (**b**) shows the Flat case where the optical axis of the receiver is parallel to the floor.

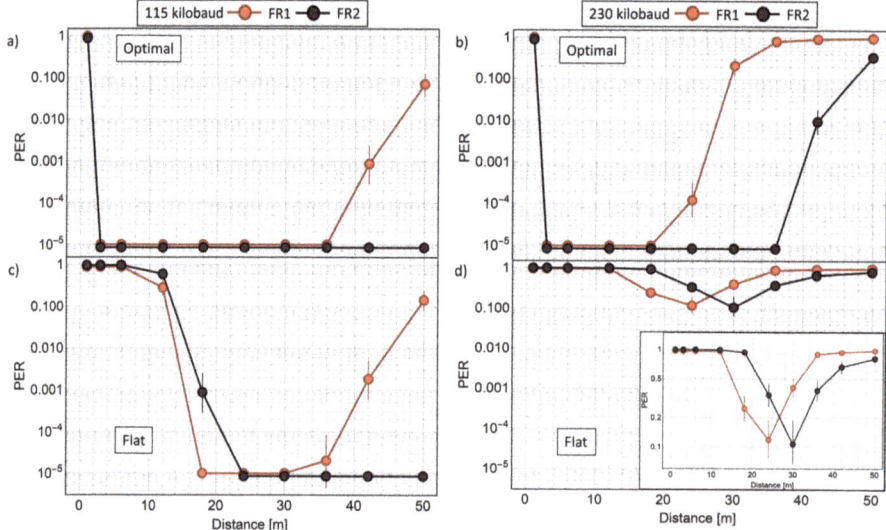

Figure 4. PER vs Distance. Two baud rates of 115 kbaud (left panel) and 230 kbaud (right panel) are used for data transmission. The sub-figures (**a**,**b**) show the Optimal configuration for two Fresnel lenses $FR1$ and $FR2$, while (**c**,**d**) represent the Flat configuration. The inset in (**d**) reports data on a zoomed vertical scale.

For performance analysis, the acceptable PER is set to 10^{-3} as recommended by most of the telecommunication applications. Therefore, PER performance for both the lenses is investigated against this threshold. It could be observed from the figure that in Optimal configuration, for both 115 and 230 kbaud, $FR2$ outperforms $FR1$ and could achieve a PER = 10^{-3} till 50 m (115 kbaud) and 40 m (230 kbaud). At the same baudrates, $FR1$ attains such performances up to 42 and 26 m, respectively. In the Flat case, for 115 kbaud, a $PER = 10^{-3}$ is achieved in a range of 18–50 and 16–41 m for $FR2$ and $FR1$, respectively. However, for 230 kbaud, the required PER is unattainable for both lenses and all distances. Our analysis shows that for our Fresnel-lens-based prototype, the most critical region not only corresponds to the far distant one, but also to the very near space surrounding the traffic light. This behavior, which is especially evident in panels c–d where the Flat case is reported, is intimately connected to the large angles involved, requiring very large FOVs, but also to the intrinsic intensity map projected by the traffic light lamp, which conveys very small intensities at short distances [29].

The designed prototype is further tested for both continuous information broadcast (Figure 5) and beaconing (Figure 6), which makes it suitable for both situational information exchange, for example, traffic congestion and alternative routes information, and continuous message broadcast in emergency events, for example, road accidents, incoming rescue vehicles, and bridge breakdowns. The non-trivial difference among the broadcast and beaconing modes, from the optoelectronic point of view, is that the latter, differently from the former, does not grant a constant RMS value of the detected signal, so that the detection of isolated bursts is typically harder due to the presence of transients in the signal, which can be hardly filtered by the AC-coupling of the RX stage. So, the capability of our system to handle both transmission modes represents a key, non-trivial feature, which could be essential in the deployment of VLC technology in realistic ITS applications.

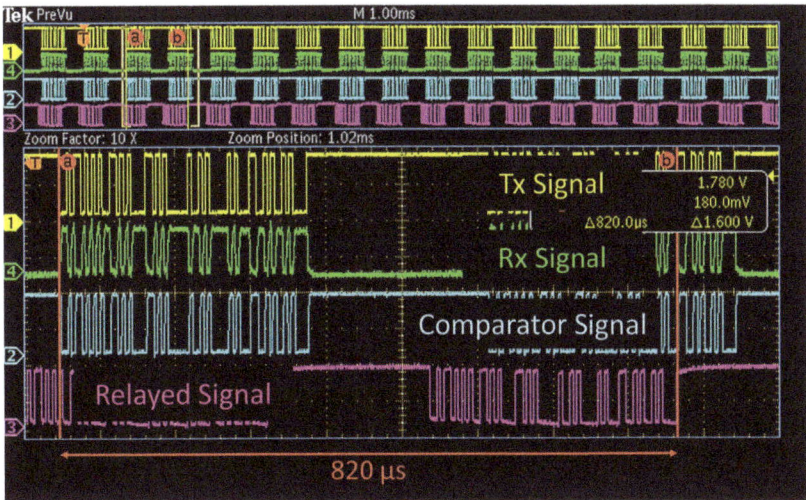

Figure 5. Information broadcast: oscilloscope screen showing the repeatedly transmitted signal (yellow), received signal (green), digitized signal (blue) and relayed signal (purple). The observed latency is 820 microseconds at 230 kbaud for the full Active Decode and Relay (ADR) process.

As clearly appearing from our characterization, a very promising aspect of this Fresnel-lens based prototype, besides what has been reported in ref. [13] with different, more expensive aspheric condensers, is the capability to reliably deliver digital optical information up to 50 m in ITS applications despite the lighter, thinner and more economic optical condenser, yet granting ultra-low latency, that is measured as low as 820 µs for 230 kbauds and 1.5 ms for 115 kbaud. Noticeably enough, our Fresnel lens-based VLC chain achieves very low latencies, even lower than those requested by IEEE802.11p standard and newly introduced C-V2X [30]. Even if our latency values are relative to the final I2V branch of the ITS network and do not consider the core network segment, it is important to remark that latencies of the final segment are relevant in case of ad-hoc networks: whilst the directionality of the VLC channel avoids the need for complex handshaking protocols between an incoming vehicle and the specific infrastructure that is casting the information, latency times of the final segment could reach very large values in RF-based networks, much beyond the end-to-end latency values given in ref. [30].

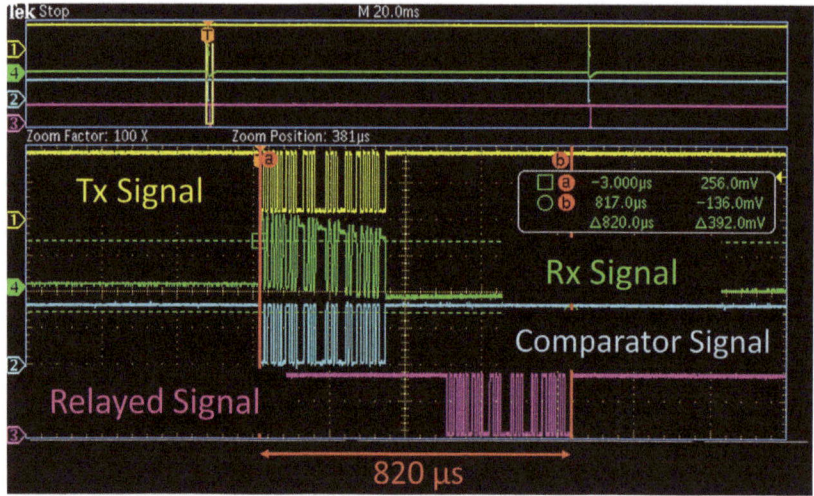

Figure 6. Beaconing: oscilloscope screen showing transmitted signal (yellow), received signal (green), digitized signal (blue) and relayed signal (purple). The beaconing interval is set much higher than the size of the packet.

4. Conclusions

In this work, we designed and tested a low-cost, low-latency VLC prototype for I2V communications, using a commercial LED-based traffic light as transmitter and a conventional amplified photodiode with low-cost Fresnel lenses as a receiver. Our VLC-based system is based on an open-source microcontroller platform (Arduino Due) and is capable of Active Decode and Relay (ADR) of received information to a further modulation stage to propagate the information to incoming units through, for example, rear car lamps.

We evaluated the PER performance of the system in various configurations for both the 1″ and 2″ Fresnel lenses. Two baud rates of 115 and 230 kbaud are used for data transmissions. When embedding a 2″ Fresnel lens in Optimal configuration, our system attains an error-free transmission (PER $< 10^{-5}$) up to 36 m with sub-millisecond (820 µs) latency of the full ADR process at 230 kbaud (50 m for 230 kbaud). For a more realistic Flat system configuration, simulating a car approaching the traffic light without an optical tracking mechanism, the error-free transmission is achieved at 115 kbaud with 2″ lens and above 24 m. If only PER = 10^{-3} is required, instead, the 18–50 m range can be covered. We also tested the VLC Fresnel-based system for both the beaconing of situational information and event-triggered message broadcast, finding it suitable for both modes of transmission. The noticeable low-latency feature of this prototype makes it integrable with current 5G-based C-V2X Intelligent Transportation Systems.

For the future, we are planning outdoor measurement campaigns for both I2V, V2V and Infrastructure-to-Vehicle-to-Vehicle (I2V2V) communications. In addition, our system will be tested for various environmental conditions.

Author Contributions: Conceptualization, T.N., M.S., S.C., L.M. and J.C.; methodology, T.N., M.S., S.C., L.M. and J.C.; software, T.N. and S.C.; validation, L.M. and J.C.; formal analysis, L.M. and S.C.; investigation, T.N., M.S. and J.C.; resources, J.C.; data curation, M.S. and S.C.; original draft preparation, T.N. and J.C.; supervision, J.C.; funding acquisition, L.M. and J.C. All authors have read and agreed to the published version of the manuscript.

Funding: This work has been carried out under the financial support of Project PON MIUR 2017 "OK-INSAID", of Progetto Premiale MIUR FOE 2015 "OpenLab", and by Project MISE "5G City"—Prato.

Acknowledgments: Authors would like to thank the Company ILES srl in Prato (Italy) for providing the traffic light and for support, and Francesco Cataliotti for precious discussions.

Conflicts of Interest: The authors declare no conflict of interest.

References

1. Karagiannis, G.; Altintas, O.; Ekici, E.; Heijenk, G.; Jarupan, B.; Lin, K.; Weil, T. Vehicular Networking: A Survey and Tutorial on Requirements, Architectures, Challenges, Standards and Solutions. *IEEE Commun. Surv. Tutor.* **2011**, *13*, 584–616. [CrossRef]
2. Kenney, J.B. Dedicated Short-Range Communications (DSRC) Standards in the United States. *Proc. IEEE* **2011**, *99*, 1162–1182. [CrossRef]
3. Cailean, A.; Cagneau, B.; Chassagne, L.; Popa, V.; Dimian, M. A survey on the usage of DSRC and VLC in communication-based vehicle safety applications. In Proceedings of the 2014 IEEE 21st Symposium on Communications and Vehicular Technology in the Benelux (SCVT), Delft, The Netherlands, 10 November 2014; pp. 69–74. [CrossRef]
4. Aa, V. *IEEE Std 802.11 p-2010, Amendment 6: Wireless Access in Vehicular Environments*; IEEE Computer Society: Columbia, DC, USA, 2010.
5. Ahmed-Zaid, F.; Bai, F.; Bai, S.; Basnayake, C.; Bellur, B.; Brovold, S.; Brown, G.; Caminiti, L.; Cunningham, D.; Elzein, H.; et al. *Vehicle Safety Communications—Applications VSC-A Second Annual Report January 1, 2008 through December 31, 2008*; Technical Report; NHSTA: Columbia, DC, USA, 2011.
6. Molina-Masegosa, R.; Gozalvez, J.; Sepulcre, M. Configuration of the C-V2X Mode 4 Sidelink PC5 Interface for Vehicular Communication. In Proceedings of the 2018 14th International Conference on Mobile Ad-Hoc and Sensor Networks (MSN), Shenyang, China, 6–8 December 2018; pp. 43–48. [CrossRef]
7. Araniti, G.; Campolo, C.; Condoluci, M.; Iera, A.; Molinaro, A. LTE for vehicular networking: A survey. *IEEE Commun. Mag.* **2013**, *51*, 148–157. [CrossRef]
8. Molina-Masegosa, R.; Gozalvez, J. LTE-V for Sidelink 5G V2X Vehicular Communications: A New 5G Technology for Short-Range Vehicle-to-Everything Communications. *IEEE Veh. Technol. Mag.* **2017**, *12*, 30–39. [CrossRef]
9. Tanaka, Y.; Haruyama, S.; Nakagawa, M. Wireless optical transmissions with white colored LED for wireless home links. In Proceedings of the 11th IEEE International Symposium on Personal Indoor and Mobile Radio Communications, London, UK, 18–21 September 2000; Volume 2, pp. 1325–1329. [CrossRef]
10. Zafar, F.; Bakaul, M.; Parthiban, R. Laser-Diode-Based Visible Light Communication: Toward Gigabit Class Communication. *IEEE Commun. Mag.* **2017**, *55*, 144–151. [CrossRef]
11. Lee, C.; Shen, C.; Cozzan, C.; Farrell, R.M.; Speck, J.S.; Nakamura, S.; Ooi, B.S.; DenBaars, S.P. Gigabit-per-second white light-based visible light communication using near-ultraviolet laser diode and red-, green-, and blue-emitting phosphors. *Opt. Express* **2017**, *25*, 17480–17487. [CrossRef] [PubMed]
12. Arnon, S.; Barry, J.; Karagiannidis, G.; Schober, R.; Uysal, M. *Advanced Optical Wireless Communication Systems*; Cambridge University Press: Cambridge, UK, 2012.
13. Nawaz, T.; Seminara, M.; Caputo, S.; Mucchi, L.; Cataliotti, F.S.; Catani, J. IEEE 802.15.7-Compliant Ultra-Low Latency Relaying VLC System for Safety-Critical ITS. *IEEE Trans. Veh. Technol.* **2019**, *68*, 12040–12051. [CrossRef]
14. Takai, I.; Ito, S.; Yasutomi, K.; Kagawa, K.; Andoh, M.; Kawahito, S. LED and CMOS image sensor based optical wireless communication system for automotive applications. *IEEE Photonics J.* **2013**, *5*, 6801418. [CrossRef]
15. Okada, S.; Yendo, T.; Yamazato, T.; Fujii, T.; Tanimoto, M.; Kimura, Y. On-vehicle receiver for distant visible light road-to-vehicle communication. In Proceedings of the 2009 IEEE Intelligent Vehicles Symposium, Xi'an, China, 3–5 June 2009; pp. 1033–1038.

16. Takai, I.; Harada, T.; Andoh, M.; Yasutomi, K.; Kagawa, K.; Kawahito, S. Optical vehicle-to-vehicle communication system using LED transmitter and camera receiver. *IEEE Photonics J.* **2014**, *6*, 1–14. [CrossRef]
17. Yamazato, T.; Takai, I.; Okada, H.; Fujii, T.; Yendo, T.; Arai, S.; Andoh, M.; Harada, T.; Yasutomi, K.; Kagawa, K.; et al. Image-sensor-based visible light communication for automotive applications. *IEEE Commun. Mag.* **2014**, *52*, 88–97. [CrossRef]
18. Goto, Y.; Takai, I.; Yamazato, T.; Okada, H.; Fujii, T.; Kawahito, S.; Arai, S.; Yendo, T.; Kamakura, K. A new automotive VLC system using optical communication image sensor. *IEEE Photonics J.* **2016**, *8*, 1–17. [CrossRef]
19. Cailean, A.; Cagneau, B.; Chassagne, L.; Topsu, S.; Alayli, Y.; Blosseville, J.M. Visible light communications: Application to cooperation between vehicles and road infrastructures. In Proceedings of the 2012 IEEE Intelligent Vehicles Symposium (IV), Alcala de Henares, Spain, 3–7 June 2012; pp. 1055–1059.
20. Corsini, R.; Pelliccia, R.; Cossu, G.; Khalid, A.M.; Ghibaudi, M.; Petracca, M.; Pagano, P.; Ciaramella, E. Free space optical communication in the visible bandwidth for V2V safety critical protocols. In Proceedings of the 2012 8th International Wireless Communications and Mobile Computing Conference (IWCMC), Limassol, Cyprus, 27–31 August 2012; pp. 1097–1102.
21. Terra, D.; Kumar, N.; Lourenço, N.; Alves, L.N.; Aguiar, R.L. Design, development and performance analysis of DSSS-based transceiver for VLC. In Proceedings of the 2011 IEEE EUROCON-International Conference on Computer as a Tool (EUROCON), Lisbon, Portugal, 27–29 April 2011; pp. 1–4.
22. Lourenço, N.; Terra, D.; Kumar, N.; Alves, L.N.; Aguiar, R.L. Visible light communication system for outdoor applications. In Proceedings of the 2012 8th International Symposium on Communication Systems, Networks & Digital Signal Processing (CSNDSP), Poznan, Poland, 18–20 July 2012; pp. 1–6.
23. Kumar, N.; Lourenço, N.; Terra, D.; Alves, L.N.; Aguiar, R.L. Visible light communications in intelligent transportation systems. In Proceedings of the Intelligent Vehicles Symposium, Alcala de Henares, Spain, 3–7 June 2012; pp. 748–753.
24. Cailean, A.M.; Cagneau, B.; Chassagne, L.; Topsu, S.; Alayli, Y.; Dimian, M. Visible light communications cooperative architecture for the intelligent transportation system. In Proceedings of the IEEE 20th IEEE Symposium on Communications and Vehicular Technology in the Benelux (SCVT), Namur, Belgium, 21 November 2013; pp. 1–5.
25. Kim, Y.H.; Cahyadi, W.A.; Chung, Y.H. Experimental Demonstration of VLC-Based Vehicle-to-Vehicle Communications Under Fog Conditions. *IEEE Photonics J.* **2015**, *7*, 1–9. [CrossRef]
26. Kirrbach, R.; Faulwaßer, M.; Jakob, B. Non-rotationally Symmetric Freeform Fresnel-Lenses for Arbitrary Shaped Li-Fi Communication Channels. In Proceedings of the 2019 Global LIFI Congress (GLC), Paris, France, 12–13 June 2019; pp. 1–6.
27. Kwon, J.K. Inverse Source Coding for Dimming in Visible Light Communications Using NRZ-OOK on Reliable Links. *IEEE Photonics Technol. Lett.* **2010**, *22*, 1455–1457. [CrossRef]
28. Khalili, R.; Salamatian, K. A new analytic approach to evaluation of packet error rate in wireless networks. In Proceedings of the 3rd Annual Communication Networks and Services Research Conference (CNSR'05), Halifax, NS, CA, 16–18 May 2005; pp. 333–338.
29. Caputo, S.; Mucchi, L.; Cataliotti, F.S.; Seminara, M.; Nawaz, T.; Catani, J. Measurement-based VLC channel characterization for I2V communications in a real urban scenario. *arXiv* **2019**, arXiv:1905.05019.
30. Emara, M.; Filippou, M.C.; Sabella, D. MEC-Assisted End-to-End Latency Evaluations for C-V2X Communications. In Proceedings of the 2018 European Conference on Networks and Communications (EuCNC), Ljubljana, Slovenia, 18–21 June 2018; pp. 1–9. [CrossRef]

© 2020 by the authors. Licensee MDPI, Basel, Switzerland. This article is an open access article distributed under the terms and conditions of the Creative Commons Attribution (CC BY) license (http://creativecommons.org/licenses/by/4.0/).

Article

Experimental Measurements of a Joint 5G-VLC Communication for Future Vehicular Networks

Dania Marabissi [1], Lorenzo Mucchi [1,*], Stefano Caputo [1], Francesca Nizzi [1], Tommaso Pecorella [1], Romano Fantacci [1], Tassadaq Nawaz [2,3], Marco Seminara [3,4] and Jacopo Catani [3,4]

1. Department of Information Engineering, University of Florence, 50139 Firenze, Italy; dania.marabissi@unifi.it (D.M.); stefano.caputo@unifi.it (S.C.); francesca.nizzi@unifi.it (F.N.); tommaso.pecorella@unifi.it (T.P.); romano.fantacci@unifi.it (R.F.)
2. Department of Physics and Astronomy, University of Florence, 50019 Sesto Fiorentino, Italy; nawaz@lens.unifi.it
3. European Laboratory of Non Linear Spectroscopy (LENS), 50019 Sesto Fiorentino, Italy; seminara@lens.unifi.it (M.S.); jacopo.catani@ino.cnr.it (J.C.)
4. National Institute of Optics (INO-CNR), 50019 Sesto Fiorentino, Italy
* Correspondence: lorenzo.mucchi@unifi.it; Tel.: +39-055-275-8539

Received: 5 June 2020; Accepted: 30 June 2020; Published: 2 July 2020

Abstract: One of the main revolutionary features of 5G networks is the ultra-low latency that will enable new services such as those for the future smart vehicles. The 5G technology will be able to support extreme-low latency thanks to new technologies and the wide flexible architecture that integrates new spectra and access technologies. In particular, visible light communication (VLC) is envisaged as a very promising technology for vehicular communications, since the information provided can flow by using the lights (as traffic-lights and car lights). This paper describes one of the first experiments on the joint use of 5G and VLC networks to provide real-time information to cars. The applications span from road safety to emergency alarm.

Keywords: visible light communications; 5G networks; smart vehicles; field trials

1. Introduction

In recent decades, the continuous increase of capacity demand resulting from massive data growth has necessitated rapid changes in wireless communication networks. Nowadays, this evolutionary trend is moving toward the fifth generation (5G) of wireless systems. Hence, 5G is envisioned to have a significant technology gap compared to previous cellular networks: very high data rate, extremely low latency, high cell capacity, massive number of connected devices, guaranteeing energy and cost-efficiency. To face these challenges, 5G networks are expected to deploy a high number of cells and to include additional spectrum with respect to current systems. Consequently, communications in millimeter-wave (mmWave) spectrum, with short-range and large bandwidth availability, are considered a key 5G-technology. Also, other spectra and technologies are under investigation. In particular, visible light communication (VLC) is considered a promising complementary technology to mmWave for short-range communication scenarios, especially for indoor and hot-spot connections [1]. VLCs can provide very high data rates, low-energy consumption, low latency, and low implementation costs. The Institute of Electrical and Electronics Engineers (IEEE) has standardized the physical layer (PHY) and medium access control (MAC) sublayer for short-range optical wireless communications (OWC), including VLC and optical camera communications (OCC) [2]. Consequently, while the 5G standardization process is ongoing, the integration of VLC in 5G systems is under investigation [3–6]. In [3,4], hypotheses on the integration between 5G and VLC are envisioned. In [5], the integration of a

VLC segment with 5G backhaul is investigated, while an overall general description of the benefits that the VLC can bring to 5G networks is given in [6,7].

Apart from the research and standardization processes, launching a new technology requires field trials to define and test the key performance indicators (KPIs) as well as to determine how the users can maximize the exploitation of the new capabilities of the 5G network. Different field trials and tests have been carried out all around the world and are currently on going. Anyway, all these tests are mainly focused on the 5G new radio (NR) technologies: massive multiple input multiple output (mMIMO), 3D-beamforming, carrier aggregation, use of new air interfaces, and advanced solutions for the core network [8,9]. In contrast, the integration of 5G NR and VLC is not largely tested, only few papers present experimental results. In [10], a performance evaluation is presented where the 5G NR air frame is adapted to the VLC transmission, while in [11], resource allocation in VLC networks is discussed.

This paper presents a significant contribute in this context. In fact, it focuses on a field-test that integrates 5G communication capabilities with VLCs in a vehicular scenario. The aim is mainly to evaluate performance that can be actually achieved. To the best of our knowledge this is the first case presenting an effective VLC_5G integrated field-trial in this application scenario. The main goal is to test the end-to-end latency of the communication that can be offered by this integrated network. The Italian Ministry of Economic Development (MiSE) started a 5G field-trial project at the end of 2017 in the cities of Prato e L'Aquila. The entire experimental project lasts four years and includes a pre-commercial experimentation during the last year. Two telecommunications operators, WindTre and OpenFiber, are the coordinators of the project since they both own significant network infrastructure, including both wireless and optical fibers. This paper first describes the main characteristics of the field-trial environment and then details the experiment carried out to deliver road information to vehicles. Information collected by sensors connected to the 5G network, is distributed by the traffic light with near-zero latency thanks to the VLC. This paper is an extension of two previous papers [12,13]. In particular, [12] presents a general description of the 5G project framework that is here recalled only on its main aspects, while [13] presents the description of the smart mobility use case. This paper focuses on the same use case but provides KPIs measurements and analysis of the performance of the integration between the 5G network and a VLC-based V2I communication system, achieved at the end of the 5G project. In addition, we have included the distribution of the empirical data of the latency for the 5G link and for the overall system (5G link and VLC link), as well as the best fitting distribution, derived by using the Bayesian information criterion (BIC). The probability density function (PDF) of the empirical data and fitting PDFs are reported. Also, cumulative distribution function (CDF) and CDF error functions are shown in the paper.

The rest of the paper is organized as follows. Section 2 describes the overall system model of the integration of the VLC system into the 5G network. Section 3 briefly highlights the 5G testing network, the use cases and the field trials. Section 4 shows the details of the VLC system, while Section 5 reports and discusses the experimental results. Section 6 concludes the paper.

2. VLC-5G Integration System Model

Before describing the 5G field trial, in this section we present the system model of the VLC-5G integration used in the field-trial that will be detailed later. The model is presented in Figure 1. In particular, the VLC segment is provided by the traffic lights of a (smart) city, while the 5G network provides the connectivity to the traffic lights.

Figure 1. VLC-5G network topology.

Let us assume that a set $\{1, \ldots, s, \ldots, S\}$ of traffic lights are serving a set $\{1, \ldots, u, \ldots, U\}$ of vehicular users. Each traffic light is connected to 5G network and it is capable to generate a VLC signal towards the incoming vehicles. Each traffic light is assumed to be within the coverage of a 5G cell. Vehicles are assumed to be equipped with a VLC receiver. Since each traffic light has a directive beam, we assume that traffic lights do not interfere each other, i.e., the VLC bandwidth can be reused.

The signal-to-noise ratio (SNR) for the uth user served by the sth traffic light through a VLC link can be written as [14].

$$\gamma_{u,s} = \frac{R_{PD} H_{u,s} P_s}{\xi} \quad (1)$$

where R_{PD} denotes the responsitivity of the photodiode, $H_{u,s}$ is the line-of-sight (LOS) VLC channel gain between the sth traffic light and uth vehicle, P_s is the transmission power, and ξ denotes the cumulative noise power. The channel gain $H_{u,s}$ depends on several variables: photodetector area, angle of irradiance, angle of incidence, signal transmission coefficient of the optical filter, refractive index of the optical concentrator, user's field of view (FOV), and the distance d between the traffic light and the user vehicle.

3. 5G Field Trials

3.1. Field-Trials Organization

The 5G field-trial project has been organized in phases and follows the standardization process. The 3rd Generation Partnership Project (3GPP) has completed the first version of the 5G standard ready for deployment in 2017. It is named non-standalone 5G NR (NSA 5G NR) and it requires a fall back to long term evolution (LTE) networks for partial operation. Currently, the stand alone (SA) 5G network architecture is under definition (to be completed in June 2020).

Project organization:

(1) Set-up phase: The ZTE Research and Innovation Lab has investigated and tested new 5G technologies.
(2) Roll-out phase: Several NR base stations (BSs) have been deployed to test all network elements and basic network functions. Initially, the deployment relays on the existing LTE-Core Network (CN) following the NSA network architecture. At the end of the project the SA network architecture (with a 5G-CN) will be deployed and evaluated.
(3) Service phase: Extensive field trials to validate the KPIs and to test innovative services provided on the 5G network infrastructure.

3.2. 5G Network

3.2.1. Network Architecture

The 5G network architecture deployed for the test-bed is based on a network slicing approach.

Network slicing is considered one of the pillars of 5G systems. Specific functions can be designed to create and manage dedicated end-to-end logical networks, without losing the economies of scale of a common physical infrastructure. Each logical network is tailored to provide a specific service and/or provide a particular tenant with a given level of guaranteed network resources [15]. Consequently, 5G systems can support a wide variety of vertical markets that originate a wide range of services.

The 5G network architecture considered in the field-test has a logical organization on three-layers (Figure 2). These are used for diversification of the service layer from the network functions and the physical infrastructure. Physical infrastructure is the first layer, whose main objective is the managing of the physical resources. Network functions is the second layer, whose scope is function customization. Service is the third layer and it maps the service level agreements (SLAs), quality of service (QoS), and required functionality into the slice's configurations. An orchestrator manages the three layers by mapping the resources available at different layers to the slices.

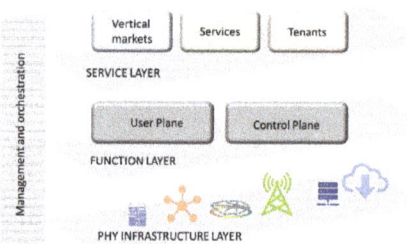

Figure 2. Network slicing conceptual architecture.

3.2.2. Physical Infrastructure

In the deployed infrastructure the cloud-radio access network (RAN) approach has been adopted. BSs are composed by several radio frequency elements called (active antenna units—AAU) and the base band unit (BBU) that is the element of processing. The BBU can ben centralized in a single point or virtualized in the cloud and represents the smart-element of the BS while only simple RF equipment are needed at the network edge.

AAUs are able to work on different frequencies, thus a multimode network is deployed. In particular, a heterogeneous multi-layer 5G cellular architecture is considered where cells of different coverage areas are overlapped and provide services in different frequency bands (e.g., 3.7 GHz and mmWave) and with different access technologies (e.g., 5G-NR, WiFi and VLC). In this case, a dual-connectivity approach is adopted. The 5G-NR AAU operating in low frequency bands provides basic services on a wide area, while small cells operating on mmWave and visible light spectra provide high data-rate services indoor and in hot-spots.

AAUs and BBU are connected though the fronthaul link that is characterized by low-latency and high-speed thanks to the adoption of the common public radio interface (CPRI) [16]. CPRI is an interface that defines the transmission of digital-radio over fiber and allows a transmission of data with a fixed bit rate over a dedicated channel.

A single BBU and a single AAU connected to a channel emulator and to a basic User Equipment (UE) were used during the very initial tests. Successively, several BBUs and AAUs, connected to the LTE-CN following the 3GPP NSA deployment scenario Options 3 and 3a [17], have been deployed in the two cities. In particular, the LTE radio and CN were used as an anchor for mobility management and coverage, while adding new 5G carrier. Hence, the LTE-RAN connects the LTE-CN with the 5G NR.

The FlexE [18] standard was used to manage via software the transport network. This allowed a flexible reconfiguration of the network, making the physical layer transparent to the service layer. 100G Ethernet rings provided connections to BBUs, while optical links provided connections with the core network.

NSA deployment allows to validate KPIs mainly related to control plane and user plane latency, user and cells peak data rates, and to test network slicing approach [15].

The SA 5G NR architecture will be deployed during the last part of the roll-out phase. In particular, 3GPP Option 2 [19], i.e., 5G NR devices are directly connected to the new 5G CN, is considered in the deployment of the network. This solution is independent on 4G network deployments and it provides simpler implementation. Anyway, the above solution requests the 5G end-to-end network to be completely defined before the pre-commercial phase, and substantial investments to provide the service coverage over all the territory. In this phase, KPIs related to mobility and handover will be tested.

The 5G NR improved air interface is based on:

- Operations in multiple frequency bands;
- Massive MIMO (mMIMO) techniques;
- Non-orthogonal multiple access (NOMA);
- Dense network deployment.

3.3. Use Cases

The project aims to validate the 5G and its role of new digital economy creators. In fact, 5G is does not provide only the enhancement of current wireless systems, but it also represents a framework where new services and applications will be provided. Consequently, 5G can support a different vertical market that need of a wide range of services each one characterized by specific requirements [19].

In this context, the project defines several use cases to test the provisioning on the 5G network of innovative services that are under deployment.

A brief description is provided in Table 1, while the use case on smart mobility is detailed in the next section.

Table 1. 5G project Use Cases.

Use Case	Description
e-Health	A platform that provides personalized care and assistance with guaranteed quality of service and continuity for telemedicine, telemonitoring and analysis of behavioural habits.
Smart industry	A digital platform to provide Industry 4.0 services for the optimization of production processes, energy efficiency, maintenance and operation.
Smart grid	A management architecture inspired by blockchain protocol enabling new services and management methods of the load and generation assets.
IoT and sensors	Connected sensors (following the IoT paradigm) for real-time remote control of the industrial processes, heavy machinery in hazardous environments, logistics optimization and products tracking.
Structural health monitoring	A monitoring service for buildings/infrastructures, reporting any anomaly of the most significant structural parameters even in emergency (e.g., earthquake) by means the use of sensors and drones.
Virtual reality for cultural heritage	An immersive virtual visit of different type of cultural heritage with digital contents delivery by using the virtual reality and the augmented reality.
Agriculture 2.0	Support and improvement of the *Made in Italy* brand. Tracking of products and production processes in the Agro-Food sector.

More details on the overall 5G test-bed can be found in [12].

Smart Mobility Use Case

This section focuses on smart mobility use case that is the one for which we present the testbed and experimental results.

It is a widely shared policy perspective that smart mobility will address numerous social challenges about freight and passenger transport: safety, efficiency, energy saving, reliability, etc. In this context, 5G is envisaged as the ultimate technology for smart mobility, since features of cellular-vehicle-to-everything (C-V2X) will be included as part of the cellular chipsets embedded into vehicles for their vehicle-to-network (V2N) communications. The vertical domain of smart mobility is therefore one of the primary driving sectors for progression towards 5G, and one of the most important use cases in the 5G project.

The smart mobility platform has two main goals:

(1) Road monitoring: electric parking and charging points are being deployed for the purpose of monitoring the state of the road surface (presence of gaps, slope, traffic conditions, etc.) during regular everyday activities by installing in the vehicles a blackbox containing a 5G module for real-time transmission of information to a data processing center. Electric cars are equipped with a differential GPS that can map the geographic positions of the holes found during regular vehicle use with a precision of cm.

(2) Advanced viability: vehicles share data with other vehicles and with a control center where data traffic information is smartly combined other information such as the city's temperature, the road status and other sensors information. The goal is to use real-time information for increasing car and driver health, comfort, and style of driving and for minimizing road traffic, congestion, and consequent emissions.

The Smart Mobility platform is based on some key technologies:

(a) Network Slicing: The 5G network is able to provide specific network slice for V2X communications in order to manage its own features independently on the other services. However, how slices can efficiently share the resources is still a challenging issue. The studying of practical algorithms is ongoing considering both the computational complexity and the ability to reconfigure the resources allocation following the variability of the vehicular network topology. In particular, one of the main challenges of the infrastructure layer is the virtualization and division of the RAN into slices due to spectrum limitation. In addition, the coexistence communications with the network (V2N) and among vehicles requires a high flexibility and dynamicity of the RAN.

(b) MEC: Reduced network congestion and improved applications performance can be obtained by using the multi-access edge computing (MEC) paradigm, which introduces cloud-computing capabilities closer to the end-user within the access network. Data generated from vehicles and infrastructure can be efficiently processed by the MEC thus delivering locally-relevant contents to support smart driving services. The MEC allows ultra-low latency, high bandwidth and real-time access to the access network that can be leveraged by the applications.

(c) Access point densification: Network capacity can be improved by deploying a large number of small cells in addition to traditional macrocells. Moreover, in case of emergency or network unavailability, vehicles themselves could complement the public network becoming moving cells. Anyway, in case of coexistence of multiple cell-layers, a careful investigation on resource usage is required as well as on coordination strategies among all the cells.

(d) Multi RATs: In the smart mobility paradigm, multiple radio access technologies (Multi-RAT) can be integrated into vehicles, which become a powerful mobile gateway. Both V2V and V2N communications (e.g., 802.11p, LTE, C-V2X, 5G, VLC) could ask for multi-RATs integration, although an accurate managing should be done for the exploitation of benefits and limitation of their drawbacks.

In this context, one of the most original experimentations is the integration of VLC in the proposed platform for V2N and V2V communications. A measurement campaign on the integration of 5G system and VLC for vehicular communications have been carried out in the project, and it is described in the following section.

4. VLC for Vehicular Services

Vehicular networks applications are envisioned to benefit of communication opportunity given by visible light [20,21]. Visible light communications (VLC) show advantages which are not inherent in RF-based technology, e.g., huge unlicensed spectrum, robustness versus jamming, and much less interference. VLC is also a green technology since the same energy used to light or for road signalling can be used to communicate with vehicles.

A preliminary experiment has been conducted at the University of Florence for I2V communications which involves VLC (Figure 3) signals. A real traffic-light, located in a real urban road, was used as VLC transmitter. A VLC receiver was moved over a grid of points, at different distances, in front of the traffic light. The experiments provided a 200 kbps data rate up to 40 m. The prototype included a custom LED driver installed in a regular traffic light. The digital information was inserted into the light of the red lamp by the LED driver which modulated the light based on the source information bits. In particular, the intensity of the LED lamp moves from $+A$ (bit 1) to $-A$ (bit 0) around a mean value, which is the nominal intensity of the traffic-light. The lens inside the traffic-light is a regular traffic-light lens. The traffic-light was located in an urban road in the industrial part of the city. The receiver is composed by a photodiode and a lens collecting the light. A high precision oscilloscope was connected to the VLC receiver for the display and record of the incoming VLC signal. The VLC signal was recorded for 400 µs in each point of a virtual grid from 2 to 40 m along the lanes of the road. These measurements and the subsequent analysis led us to develop an accurate propagation model as well as a performance evaluation [22,23].

Figure 3. I2V measurements campaign using a VLC-based traffic-light (city of Prato).

As depicted in Figure 1, the framework envisioned to implement a 5G modem into each traffic light of the (smart) city, while each vehicle should include a VLC receiver. Computational complexity mainly depends on the type of information provided by the infrastructure to the incoming vehicles. As in large vehicular networks, information sent to vehicles or exchanged by vehicles is typically not "heavy", also due to the dynamic topology of the network. Messages to vehicles do not usually require a large amount of bits (alarms, traffic information, etc.), thus the computational complexity or the scalability of the network does not seem to be a hard task.

5. Advanced Viability Experimental Activity

5.1. Test-Bed Description

This section describes the experimental set up and results of a joint 5G-VLC networks for vehicular applications. In particular, road safety as well as emergency alarm services have been tested. The test-bed was implemented at the Polo Universitario Città di Prato (PIN). The experimental set-up scheme is reported in Figure 4.

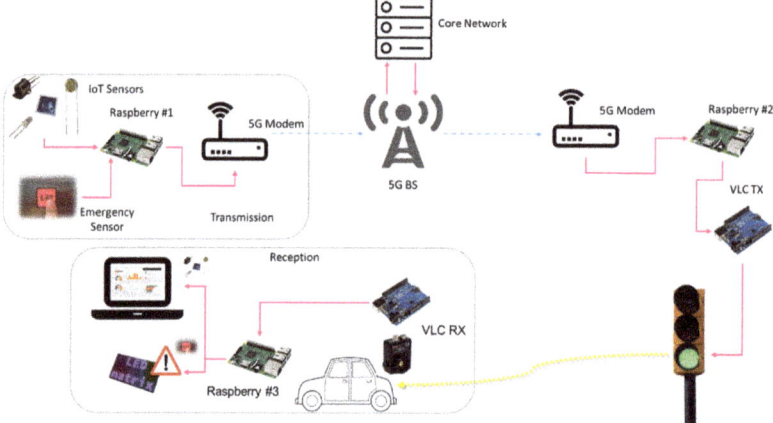

Figure 4. Test-bed architecture of the 5G-VLC joint network.

In particular, the system was composed by three Raspberry Pi 3 Model B+, 3 WeMos D1 mini, an Italian regulation compliant traffic light, a VLC transmitter, a VLC receiver, and several different IoT sensors:

- Flame sensor: To simulate a fire alarm.
- Gyroscope/accelerometer: To detect an incident between to (scale model) cars.
- Temperature, humidity and pressure sensors: To detect the presence of ice on the road.

The first two raspberries are used to collect data coming from the sensors and forward, through the 5G network, to the VLC network. The third raspberry was used only to display in the vehicle the sensors values as well as the alarm. It is important to point out that the temperature/humidity/pressure sensors send the data periodically, while the other sensors send an alarm only if an event is triggered. The raspberries are connected to the 5G customer premise equipment (CPE), provided by the 5G network operator, though ethernet gigabit ports. The traffic-light and the car were connected using custom-designed TX and RX stages based on open source Arduino device [22]. The second raspberry and the VLC block are connected by a serial cable (USB). The connection between the sensors and the raspberry is through the air: Raspberry provides a Wi-Fi Access Point and all the WeMos have an IP address. All the code in the raspberries is written in Python 3 and all the code used to drive the sensor—WeMos is written in C/C++. The VLC transmitter and receiver are IEEE802.15.7-compliant and they have been specifically designed/prototyped for I2V communications (see [22] for all hardware implementation details). The implementation design of the transmitter includes on-off keying (OOK) modulation with Manchester encoding, while the receiver collects the light by using an aspherical two uncoated lens and it is designed to reject the sun light. The most expensive part is the photodiode. Although the initial cost of the VLC prototype could be considered high (few hundreds of euros), after industrialization the cost could be reduced to few tens of euros.

5.2. Experimental Results

The performance of the overall joint network is measured by the latency metric. The results are taken by an oscilloscope connected at the first raspberry (yellow line in Figures 5 and 6), at the second Raspberry (purple line) and at the VLC receiver (green line). The yellow line, when high, represents an alarm trigger, the rising edge in the pink line represents the time needed to forward the packet on the USB cable, and the green line represents the packet arrived at the VLC receiver.

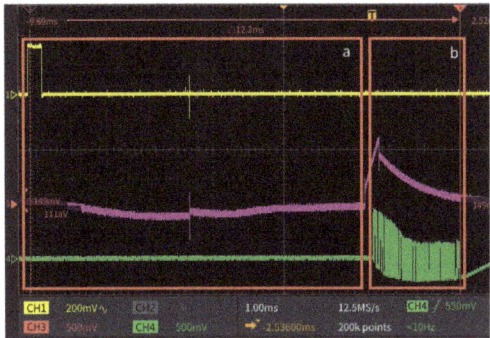

Figure 5. Maximum measured end-to-end latency.

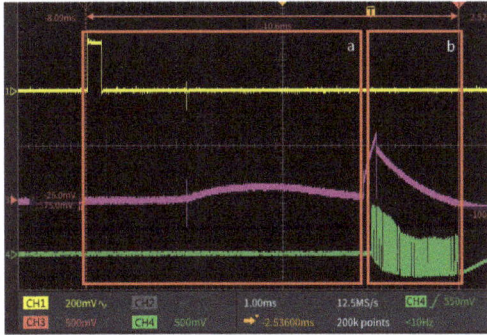

Figure 6. Minimum measured end-to-end latency.

Figures 5 and 6 show the maximum and the minimum measured end-to-end latency time. With end-to-end we intend here the time interval from the generation of one packet by the sensors to the correct reception of the packet by the VLC receiver in the car. The rectangle named a in Figures 5 and 6 shows the 5G latency time, while the rectangle named b reports the VLC latency time. The portion between the two rectangles represents the processing time introduced by the third raspberry Pi3.

Figures 7 and 8 show the distribution of the latency for the 5G segment and for the overall system (5G link and VLC link), respectively. The latency was calculated over a transmission of 2250 packets. The most frequent latency for the 5G link is about 9.5 ms, while for the VLC link is about 2.5 ms. The 5G network latency time ranges from 2.4 to 29 ms, while the VLC network latency time varies from 2.4 to 3.1 ms. This time describes the data transmission time through the optical channel at the rate of 100 kbps, as defined by the IEEE 802.15.7 standard for outdoor applications. The processing time of the second raspberry Pi 3 is highly stable, and it is equal to a few hundreds of µs.

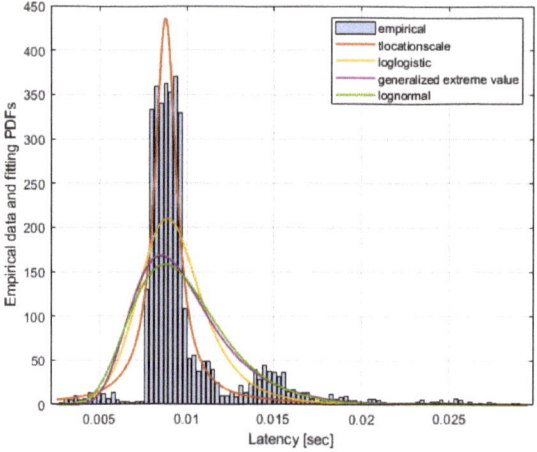

Figure 7. Empirical distribution of the latency of the 5G link only and PDFs of the fitting distributions.

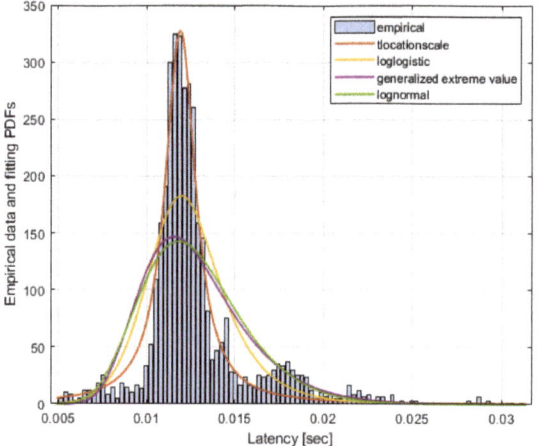

Figure 8. Empirical distribution of the latency of the overall system (5G link and VLC link) and PDFs of the fitting distributions.

The best fitting distribution of the latency for both the 5G link and overall system has been derived by using the Bayesian information criterion (BIC) [24]. Seventies different distributions have been evaluated as fitting models for the empirical data, and the distribution that minimizes the BIC was selected. For both the 5G link and the overall system, the distribution that best fits the empirical data is the t-location scale whose probability density function (PDF) $f(x|\mu, \sigma, \nu)$ is:

$$f(x|\mu, \sigma, \nu) = \frac{\Gamma\left(\frac{\nu+1}{2}\right)}{\sigma\sqrt{\nu\pi}\Gamma\left(\frac{\nu}{2}\right)}\left[\frac{\nu+\left(\frac{x-\mu}{\sigma}\right)^2}{\nu}\right]^{-\left(\frac{\nu+1}{2}\right)} \quad (2)$$

where $\Gamma(\ )$ is the gamma function, μ is the location parameter, σ is the scale parameter, and ν is the shape parameter. The best fit t-location scale parameters [μ, σ, ν] for the 5G link are [0.0088, 7.43 × 10^{-4}, 1.09], while for the overall system are [0.0119, 0.001, 1.253]. Table 2 summarizes the results.

Table 2. Best fitting distribution for 5G and VLC technologies.

Technology	Latency: Best Fitting Distribution	Distribution Parameters [μ, σ, ν]
5G	t-location scale	[0.0088, 7.43 × 10^{-4}, 1.09]
VLC	t-location scale	[0.0119, 0.001, 1.253]

Figures 9 and 10 show the cumulative distribution function (CDF) of the four best fitting distribution models for the 5G link and for the overall system, respectively. As it can be seen, the CDF error is minimized by the t-location scale distribution in both cases.

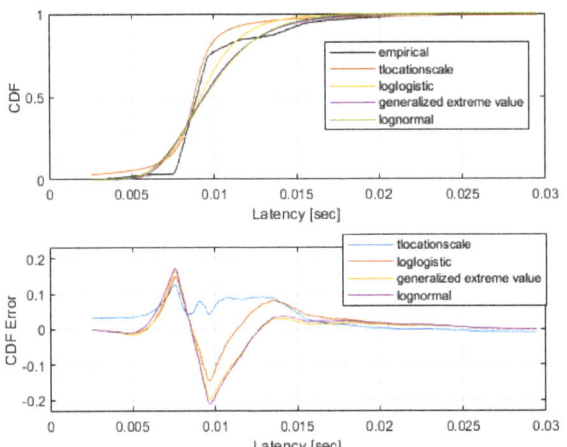

Figure 9. CDF and CDF error of the latency for empirical data and fitting distributions. 5G link only.

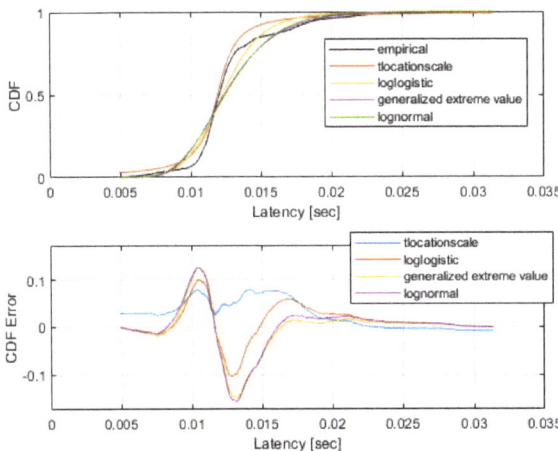

Figure 10. CDF and CDF error of the latency for empirical data and fitting distributions. Overall system (5G link and VLC link).

Although the NSA-5G network shows a very low latency compared to previous generations, we can assert that the longest part of the end-to-end latency time of the joint 5G-VLC network is introduced by the 5G part due to the packet error rate (PER) optimization. The 5G network provider has implemented techniques to reduce the PER at the 5G receiver. This means that a redundancy of the transmitted packet is introduced, during the network set up phase, to obtain PER = 10^{-6} at the receiver.

The VLC latency time could even be decreased by reducing the packet length, i.e., by optimizing the message length.

6. Conclusions

This paper introduced one of the first experiment on the joint use of a 5G and VLC network to provide information to cars. Data from road sensors has been collected by a 5G network and then sent to a VLC network for diffusion to cars through traffic-lights. In addition, data from on-demand emergency alarm has been implemented. The experiment aimed to measure the overall latency. The results showed that a total latency of about 12 ms can be reached. In particular, the most frequent latency for the 5G link was about 9.5 ms, while for the VLC link, it was about 2.5 ms. The main contribution is related to the 5G network part, that should be optimized, especially with the deployment of the SA 5G architecture.

Author Contributions: Conceptualization, D.M. and L.M.; software, S.C., T.N. and M.S.; validation, T.P. and J.C.; data curation, S.C. and F.N.; writing—original draft preparation, D.M. and L.M.; writing—review and editing, T.P. and R.F.; supervision, R.F., J.C. and L.M. All authors have read and agreed to the published version of the manuscript.

Funding: This research was carried out as part of the "5G City" project, supported by the Italian Ministry of Economic Development.

Acknowledgments: The authors would like to thank the Associate Editor and the Assistant Editor for managing the overall publication process as well as the anonymous reviewers for their useful suggestions to improve the quality of the manuscript.

Conflicts of Interest: The authors declare no conflict of interest.

References

1. Feng, L.; Hu, R.Q.; Wang, J.; Xu, P.; Qian, Y. Applying VLC in 5G Networks: Architectures and Key Technologies. *IEEE Netw.* **2016**, *30*, 77–83. [CrossRef]
2. IEEE. IEEE Standard for Local and metropolitan area networks—Part 15.7: Short-Range Optical Wireless Communications. In *IEEE Std 802.15.7-2018 (Revision of IEEE Std. 802.15.7-2011)*; Institute of Electrical and Electronics Engineers (IEEE): New York, NY, USA, 2019; pp. 1–407.
3. Rahaim, M.B.; Little, T.D.C. Toward practical integration of dual-use VLC within 5G networks. *IEEE Wirel. Commun.* **2015**, *22*, 97–103. [CrossRef]
4. Ulgen, O.; Ozmat, U.; Gunaydin, E. Hybrid Implementation of Millimeter Wave and Visible Light Communications for 5G Networks. In Proceedings of the 2018 26th Telecommunications Forum (TELFOR), Belgrade, Serbia, 20–21 November 2018; pp. 1–4.
5. Wu, S.; Wang, H.; Youn, C.-H. Visible light communications for 5G wireless networking systems: From fixed to mobile communications. *IEEE Netw.* **2014**, *28*, 41–45. [CrossRef]
6. Haas, H.; Cheng, C. Visible Light Communication in 5G. In *Key Technologies for 5G Wireless Systems*; Cambridge University Press: Cambridge, UK, 2017.
7. Haas, H. LiFi is a paradigm-shifting 5G technology. *Rev. Phys.* **2018**, *3*, 26–31. [CrossRef]
8. Shafi, M.; Molisch, A.F.; Smith, P.J.; Haustein, T.; Zhu, P.; De Silva, P.; Tufvesson, F.; Benjebbour, A.; Wunder, G. 5G: A Tutorial Overview of Standards, Trials, Challenges, Deployment, and Practice. *IEEE J. Sel. Areas Commun.* **2017**, *35*, 1201–1221. [CrossRef]
9. Ptzold, M. 5G developments are in full swing [mobile radio]. *IEEE Veh. Technol. Mag.* **2017**, *12*, 4–12. [CrossRef]
10. Shi, L.; Li, W.; Zhang, X.; Zhang, Y.; Chen, G.; Vladimirescu, A. Experimental 5G New Radio integration with VLC. In Proceedings of the 2018 25th IEEE International Conference on Electronics, Circuits and Systems (ICECS), Bordeaux, France, 9–12 December 2018; pp. 61–64.
11. Tsiropoulou, E.; Gialagkolidis, I.; Vamvakas, P.; Papavassiliou, S.; Mitton, N.; Loscrí, V.; Mouradian, A. Resource Allocation in Visible Light Communication Networks: NOMA vs OFDMA Transmission Techniques. In *Computer Vision*; Mitton, N., Loscri, V., Mouradian, A., Eds.; Springer: Berlin/Heidelberg, Germany, 2016; Volume 9724, pp. 32–46.

12. Marabissi, D.; Mucchi, L.; Fantacci, R.; Spada, M.R.; Massimiani, F.; Fratini, A.; Cau, G.; Yunpeng, J.; Fedele, L. A Real Case of Implementation of the Future 5G City. *Future Internet* **2018**, *11*, 4. [CrossRef]
13. Nizzi, F.; Nawaz, T.; Catani, J.; Seminara, M.; Pecorella, T.; Caputo, S.; Mucchi, L.; Fantacci, R.; Bastianini, M.; Cerboni, C.; et al. Data dissemination to vehicles using 5G and VLC for Smart Cities. In Proceedings of the 2019 AEIT International Annual Conference (AEIT), Florence, Italy, 18–20 September 2019.
14. Tsiropoulou, E.; Vamvakas, P.; Papavassiliou, S.; Singhal, C.; De, S.; Xu, X. Resource Allocation in Multi-Tier Femtocell and Visible-Light Heterogeneous Wireless Networks. In *Advances in Wireless Technologies and Telecommunication*; IGI Global: Hershey, PA, USA, 2017; pp. 210–246.
15. Ordonez-Lucena, J.; Ameigeiras, P.; Lopez, D.; Ramos-Munoz, J.J.; Lorca, J.; Folgueira, J. Network Slicing for 5G with SDN/NFV: Concepts, Architectures, and Challenges. *IEEE Commun. Mag.* **2017**, *55*, 80–87. [CrossRef]
16. De La Oliva, A.; Hernandez, J.A.; Larrabeiti, D.; Azcorra, A. An overview of the CPRI specification and its application to C-RAN-based LTE scenarios. *IEEE Commun. Mag.* **2016**, *54*, 152–159. [CrossRef]
17. 3rd Generation Partnership Project. TR 23.799 Technical Specification Group Services and System Aspects. In *Study on Architecture for Next Generation System Tech. Rep. V14.0.0, Dec. 2016*; European Telecommunications Standards Institute: Sophia-Antipolis, France, 2016.
18. Optical Internetworking Forum. *IA OIF-FLEXE Flex Ethernet Implementation Agreement*; Tech. Rep. 01.0; OIF: Fremont, CA, USA, 2016.
19. 5G PPP Use Cases and Performance Evaluation Models. No. 1.0. Available online: http://www.5g-ppp.eu/ (accessed on 30 June 2020).
20. Masini, B.M.; Bazzi, A.; Zanella, A. Vehicular Visible Light Networks for Urban Mobile Crowd Sensing. *Sensors* **2018**, *18*, 1177. [CrossRef] [PubMed]
21. Mucchi, L.; Cataliotti, F.S.; Ronga, L.; Caputo, S.; Marcocci, P. Experimental-based propagation model for VLC. In Proceedings of the European Conference on Networks and Communications (EuCNC), Oulu, Finland, 12–15 June 2017.
22. Nawaz, T.; Seminara, M.; Caputo, S.; Mucchi, L.; Cataliotti, F.S.; Catani, J. IEEE 802.15.7-Compliant Ultra-Low Latency Relaying VLC System for Safety-Critical ITS. *IEEE Trans. Veh. Technol.* **2019**, *68*, 12040–12051. [CrossRef]
23. Caputo, S.; Mucchi, L.; Cataliotti, F.; Catani, J. Measurement-based VLC Channel Characterization for I2V Communications in a Real Urban Scenario. *arXiv* **2019**. Available online: https://arxiv.org/abs/1905.05019 (accessed on 30 June 2020).
24. Box, G.E.P.; Jenkins, G.M.; Reinsel, G.C.; Ljung, G.M. *Time Series Analysis: Forecasting and Control*, 5th ed.; Wiley: Hoboken, NJ, USA, 2015.

 © 2020 by the authors. Licensee MDPI, Basel, Switzerland. This article is an open access article distributed under the terms and conditions of the Creative Commons Attribution (CC BY) license (http://creativecommons.org/licenses/by/4.0/).

Article
Adaptive Probabilistic Flooding for Information Hovering in VANETs

Andreas Xeros [1,†,‡], Taqwa Saeed [2,*,‡], Marios Lestas [2,‡], Maria Andreou [3,‡], Cristiano M. Silva [4,5,‡] and Andreas Pitsillides [1,‡]

1. Department of Computer Science, University of Cyprus, Nicosia 2109, Cyprus; andreas.xeros@gmail.com (A.X.); andreas.pitsillides@ucy.ac.cy (A.P.)
2. Electrical Engineering Department, Frederick University, Nicosia 1036, Cyprus; eng.lm@frederick.ac.cy
3. School of Science and Engineering of the Open University of Cyprus, Nicosia 2220, Cyprus; mandreou1@gmail.com
4. Departamento de Tecnologia, Universidade Federal de São João del-Rei (DTECH/UFSJ), Ouro Branco 36420-000, Brazil; cristiano@ufsj.edu.br
5. Departamento de Computação, Universidade Federal de Ouro Preto (DECOM/UFOP), Ouro Preto 35400-000, Brazil
* Correspondence: taquasalaheldin@gmail.com
† Nicosia 2109, Cyprus.
‡ These authors contributed equally to this work.

Received: 7 May 2020; Accepted: 4 June 2020; Published: 11 June 2020

Abstract: Information hovering is an information dissemination concept over a mobile set of peers which has not been investigated to the extent that other information dissemination paradigms have. It naturally appears in many vehicular network applications where information must be made available to vehicles within a confined geographical area for during some time period. One elementary strategy is to flood the area with data. Even in this case, some vehicles may never receive the content due to potential partitions created by low traffic density. In order to address this issue, in this work we propose a strategy based on epidemic routing in the hovering area, and probabilistic flooding outside it. Vehicles outside the hovering area serve as bridges towards partitions, leading to high reachability. We highlight the adaptive feature of the protocol, where the rebroadcast probability in partitions is adaptively regulated based on estimates of the density of vehicles in the hovering area. The performance of the proposed scheme is evaluated in VISSIM, using as the reference model in all simulation experiments a section of the road network in cities of Washington. The proposed protocol is shown to achieve the set design goals.

Keywords: vehicular networks; probabilistic flooding; vehicular communication

1. Introduction

Vehicular networks [1], despite many years of research, still constitute an area of intense research activity, as many issues remain unresolved and are further challenged by emerging technologies, such as Internet of Things (IoTs), 5G/6G networks [2], cloud computing, and so forth. Depending on the messages exchanged, vehicular ad-hoc networks (VANETs) can serve several purposes [3]. They can greatly improve safety by informing drivers of imminent road hazards [4], collision avoidance systems [5], synchronize traffic lights [6], improve comfort in driving by re-routing drivers to avoid traffic jams, monitor the driving style [7] and also serve as entertainment tools by enabling on-the-road interactive social networking [8], games, file sharing [9] and dissemination of advertisements to interested clients [10–13]. Vehicular applications typically require the exchange of messages logically attached to specific areas [14]. For instance, in the occurrence of traffic accidents, surrounding vehicles must be notified so as to take appropriate safety measures and hence avoid the critical region. Moreover, the information must lie in the area for a specific amount of time, so that joining vehicles are notified of the hazard [15,16]. A similar situation is found on commercial enterprises advertising products to possible customers. They can rely on the existing vehicular network to spread announcements in the surrounding region.

In the aforementioned situations, useful information must be broadcast to all vehicles in a specific region, and the information must be available during a predefined time period (depending on the application) in order to be noticed by vehicles. The requirements, as described above, are closely related to the more general concept of information hovering. The term 'information hovering' is becoming an established concept of information dissemination over a mobile set of peers. It was firstly introduced in Reference [17] and refined in References [18–23]. According to Reference [24], information hovering involves decoupling the hovering information from its host and coupling it to a given region called *anchor location*. Hovering information becomes attached to the specific region (called *anchor area*). Data moves from one mobile device to another trying to remain as closest as possible to the target region. Information hovering can be employed in several applications in mobile networks [21,25–27]. In the subsequent text, the terms target area, hovering area, and anchor area are used interchangeably.

One problem often encountered is the problem of low reachability within the hovering area when the network is partitioned. One solution is to allow exchange of messages outside the hovering area, originally employed in Reference [28]. This work which is based on our previous effort described in Reference [24] considers an alternative strategy based on epidemic routing within the hovering area and adaptive probabilistic flooding outside the area. Our approach allows controlled exchange of messages in the entire network offering more available paths towards the hovering area, leading to more reachability. Through simulations, we demonstrate that our approach outperforms the scheme proposed in Reference [28] by reducing the number of required messages in order to achieve high reachability. The major challenge is related to designing the rebroadcast probability function. Among several candidate functions, we select the one providing better performance, and we tune its parameters considering the phase transition phenomena (typical in probabilistic flooding schemes [29]). The proposed protocol is adaptive in the sense that the rebroadcast probability outside the hovering area is adaptively regulated based on estimates of the density of vehicles inside the hovering area. The estimates are obtained using the number of neighbors of each vehicle. The analytical formulation relating both quantities is derived using a simple model of the transportation network within the hovering area.

The results indicate that the proposed strategy satisfies the design goals and outperforms important baselines, such as epidemic routing and the scheme proposed in Reference [28]. Furthermore, we also demonstrate that in the case we use the same design methodology to render the protocol proposed in Reference [28] adaptive, its performance becomes comparable to the performance of the scheme proposed in this paper. However, our contribution in this work is more generalised, and extends to the introduction of a design procedure which can be used to design a class of density adaptive hovering protocols in other environments beyond the specific information hovering protocol. It is important to note that the universality of the proposed protocol and its applicability in areas beyond the vehicular ad hoc networking context considered in this paper, was demonstrated by the authors in Reference [30]. Therein, the proposed information hovering protocol was successfully used as a data dissemination offloading tool in a wireless data broadcasting application. The collaborative scheme between wireless broadcasting and information hovering was successful in significantly reducing the mean waiting times, thus improving the overall system performance.

An original version of this work was presented in Reference [24]. This paper extends the original work as detailed below:

- It includes a more extensive literature review, reporting recent works pertinent to the considered problem.

- It demonstrates the effectiveness of the proposed estimation algorithm by comparing the performance of the proposed protocol (which uses the estimation algorithm) with that of a modified version which assumes a known density.
- The effectiveness of the proposed methodology is demonstrated by designing a density adaptive extended area scheme which adaptively regulates an area beyond the hovering in which message exchange is allowed. This scheme is an adaptive version of the scheme proposed in Reference [28] and has not been presented in any previous work.

The remainder of this work is organized as follows. Section 2 presents the related work. Section 3 provides a formal definition of the information hovering problem in VANETs. Section 4 presents the adopted methodology leading to the development of the proposed information hovering protocol. Section 5 derives and validates the mathematical model relating the vehicle density within the hovering area and the number of neighbors of each vehicle. Section 6 evaluates the performance of the proposed scheme using simulations. Section 7 concludes the work.

2. Related Work

Information hovering deals with the persistent information dissemination of messages to all interested clients in a particular geographical area. This paradigm is highly relevant and can be realized using other information dissemination concepts as for example multi-casting, geocasting, broadcasting, beaconing and vehicular delay tolerant networks. Now, we review the aforementioned concepts and explain how our proposed approach can be placed in the presented context. A popular method for disseminating messages to interested clients is broadcasting. Despite its simplicity and high reachability, simple broadcasting (i.e., blind flooding) is avoided in highly dense networks because of the congestion and potential network meltdown it causes as a result of the high numbers of redundant messages [31]. Therefore, a number of message control schemes have been proposed in the literature to suppress the number of rebroadcasting nodes as for example: distance-based schemes, location-based schemes, neighbor knowledge schemes, counter-based schemes and probability based schemes [32]. The use of a hybrid architecture involving static and mobile antennas is presented in Reference [33].

Distance and location based methods [34–37] determine a rebroadcast according to estimates of the distance between the transmitter and the receiver. Nodes closer to the boundary of the transmitter range are chosen for retransmission, since this would lead to a larger coverage area. Distance and location based schemes both require measurements or estimates of the distance which is typically done using received signal strength as in distance based schemes or using a positioning system as in location based schemes, methods which are prone to errors and increase the implementation complexity. Neighbor knowledge or density-based schemes estimate the density of the network based on calculations of the number of neighbors to decide whether or not to retransmit [38–41]. High density is indicated by larger number of neighbors which implies that a small number of retransmission satisfies the demand of information delivery. Counter based schemes, on the other hand, monitor the number of duplicate messages received to determine a transmission [42,43]. A node receiving small number of received copies of the message is considered lying in a lightly dense area or at the boundary.

Moreover, probabilistic methods [44] where a node decides to retransmit a message based on a particular probability. The calculation of the transmission probability can be as simple as a fixed probability used by all nodes in the network, or can be based on one or more of the aforementioned parameters (i.e., distance, position, neighbor knowledge and received message statistics) [45]. Designing probabilistic methods often handles the trade-off between implementation complexity and performance. This versatility in designing probabilistic broadcasting methods, in addition to their robustness against nodes mobility and failures, renders them desirable candidates for information dissemination in wireless ad hoc networks. A combination of blind and probabilistic flooding is employed in the proposed method as discussed later in this work.

Since information hovering involves multiple nodes in a particular geographical area receiving the relevant information, multicasting and geocasting is also relevant [11,46]. Geocast protocols [47–51] disseminate the intended message among nodes in a specific area. However, the message is transmitted once and is not coupled with the reception area afterwards. This is different than the information hovering concept where the information is associated with the target area and transmitted to all nodes entering this area, which is similar in nature to the so called time-stable geocasting [52,53]. In Reference [52], authors propose geocasting methods based on unicasting the intended message to a node close to (or inside) the geocast region, and then use periodic flooding inside the region.

Many of the aforementioned methods are single-shot transmission methods intended for fully connected networks [54], which may not be always the case in vehicular networks. The high node mobility and the dynamic network topology may cause the links between nodes to frequently fail. Also, the network might be partitioned (or fragmented) due to physical obstacles, finite transmission ranges, and losses caused by congestion and interference [55]. In addition, vehicular networks can be deployed in rural areas, requiring connection in complex topographic forms, such as hills and dense forests, which may cause significant signal attenuation and reflection [46]. In fact, intermittent connectivity problem is a major challenge in the information hovering paradigm, as pointed out in Reference [18], and the proposed scheme in this work aims at addressing this problem. Persistent information dissemination has been used in literature to alleviate the problem, as for example, in methods incorporating roadside units (RSUs) as relay nodes, supplementing vehicle-to-vehicle communications with satellite communications, beaconing and Delay Tolerant Networks (DTN). Incorporating roadside units to connect sparse networks may require the deployment of a large number of units which can be costly and in some cases unfeasible [1,14]. Employing, on the other hand, satellite communications in vehicular networks does increase the connectivity [56], but it comes at a high cost.

Beaconing is the process of sending periodic short messages to direct neighbors, used by nodes in VANETs to exchange status information such as speed and location [57]. This method is relevant because of its persistent nature which ensures continuous delivery of the intended information to the targeted nodes. The exchange of beacons among nodes must be frequent enough to deliver reliable data and scarce enough not to overload or waste the network resources. One of the methods employed to achieve this balance is adaptive beaconing, which can be accomplished in different ways, as reviewed in Reference [58]. In Reference [58], authors classify adaptive beaconing approaches into message frequency control (MFC) approaches, transmit power control approaches, miscellaneous approaches and hybrid approaches which combine two or more of the former approaches. Similar to the aforementioned broadcast suppression approaches, message frequency control algorithms may also use the position and density of vehicles to determine the frequency of beaconing broadcasts [59,60]. Density adaptive beaconing is important in addressing the low connectivity challenges. In addition, message frequency control design can be highly affected by fairness considerations as these are dictated by event-driven messages which are given higher priority [61]. Transmit power control approaches (TPC), on the other hand, vary the transmission power of vehicles to alleviate congestion. The power can be regulated randomly or dictated by specific parameters, such as message fairness [62,63].

Delay Tolerant Networks are originally proposed for interplanetary communications [64], and have been adopted as a solution for MANETs with low connectivity. They have also been used for low connectivity vehicular delay tolerant networks (vDTNs) and a review is provided in Reference [55]. Similar to DTNs, regular IP networks also use store-and-forward switching. However, IP networks typically assume the existence of connected networks with reasonably short Store Times and short round trip times. On the other hand, Delay Tolerant Networks have to cope with high round trip times and network links which are not always available. Large round trip times pose specific challenges on protocol design as for example in TCP [65]. This problem has been addressed in DTNs by the so called Bundle approach [66], which collects data into bundles storing them when no links are available, and forwarding them when a communication channel is set up. This is widely known as the store-carry-and-forward approach. It is important to notice that vDTNs may have relay nodes, fixed in areas where the network is expected to be sparse, where client nodes may upload or download bundles when no other nodes are nearby [67].

A number of information dissemination concepts which have been proposed for ad hoc networks (e.g., flooding, probabilistic flooding and geographical routing) can be appropriately adapted to serve vDTNs [68]. Flooding in vDTN, can be achieved by opportunistically transmitting multiple copies of the intended bundle to different nodes. As expected, this increases the reachability of the transmission, at the cost however of potential congestion, and excessive resource utilization [69]. Another notable flooding based method is proposed in Reference [70]. Stochastic flooding in vDTN relies on random search of the destination node. For example, in First Contact scheme proposed in Reference [71] a node stochastically transmits bundles to the first node it comes in contact with. However, this random transmission may lead the bundle to sway back and forth within a set of nodes.

Moreover, geographical routing in vDTNs employs location information provided by positioning systems to choose the node that is closer to (or moving in the direction of) the destination as the next bundle forwarder. This method is receiving significant attention as means of improving the performance of routing in vDTNs [54,55], as it enables "carrying" the information towards the destination and thus enhancing the delivery rate of the store-carry-and-forward algorithm. For example, GeoSpray proposed in Reference [54] assumes prior knowledge of the position of the destination node. Then, multiple copies of the intended bundle are distributed to the nodes closer or moving towards the destination to increase the probability of delivery. It is worth mentioning that the number of transmitted copies is limited to mitigate the effects of flooding, however, transmitting more than a single copy is the main difference between GeoSpray and the previously proposed GeOpps which relies on single copy forward [72]. Other types of vDTN routing mechanisms are reviewed in Reference [68].

Limited number of works have considered information hovering as a data dissemination technique in VANETs environments [21]. Reference [73] proposes a hovering-based dissemination protocol for data transmission in road intersections. The method selects multiple one-time broadcasting nodes and only one periodically-broadcasting node at a time. The periodically broadcasting node must reside inside the hovering area and selects the new periodically broadcasting node before leaving the area. Complimentarily, Reference [23] discusses an algorithm that calculates the rebroadcasting period of the warning message. While References [73,74] target intersections in particular, Reference [23] requires the vehicle that initiates the broadcast to remain inside the hovering area, otherwise the algorithm terminates, which is not always feasible and can be unsuitable for the highly dynamic nature of VANETs. Another work that considers information hovering is Reference [75]. Vehicles inside the anchor area probabilistically rebroadcast the intended message. The broadcast probability is calculated based on the number of informed neighbors. This method requires constant update of neighbor tables which can cause a large overhead. Additional hovering-based schemes include References [76,77].

3. Problem Description

Next a formal definition of the information hovering problem in VANETs is presented.

We assume vehicles equipped with communication, information storage and processing capabilities able to form a vehicular adhoc network. Roadside units (RSU) aiming to enhance the communication may be present but are not considered here. The VANET may be represented by a graph $G(V, E)$ where V represents the set of vehicles v_i (nodes) participating in the VANET and E the set of edges with edge $l_{ij} \in E$ if vehicle v_i lies in the communication range and can exchange messages with vehicle vehicle v_j. We consider a bounded geographical area H referred to as the **hovering area** thus defining an induced subgraph $\tilde{G}(\tilde{V}, \tilde{E})$ consisting of all vehicles and their corresponding edges residing in H. It must be noted that vehicles (nodes) in G and $\tilde{G}$ are inherently mobile, abiding to any abstract mobility model thus rendering the graphs time varying. No specific mobility model is assumed. We initially assume that at least one vehicle residing in H possesses critical information. We refer to the vehicles initially possessing critical information as **information sources** with the objective being to disseminate the information to all vehicles within the hovering area. The way information sources initially generate or receive the critical information is not relevant to this problem formulation and hence it is not discussed in this work. The main objective of this study is to investigate how information sources can use the vehicular network to persistently disseminate the critical messages to all vehicles in the hovering area.

Persistent dissemination implies that the desired property of all vehicles receiving the message is not instantaneously achieved, but holds for an arbitrary time interval during which, vehicles dynamically enter and leave the hovering area. Different approaches can be adopted to solve the problem as described above. However, in our case we focus our solution space toward solutions which make efficient use of the available resources such as storage, power, bandwidth and so forth. Thus, to realise this goal, we pose the additional objective of minimizing the number of exchanged messages. By exchanging small numbers messages less storage, less power and less bandwidth may be utilized and above all a smaller latency of information delivery can be achieved, since phenomena like severe congestion and contention which lead to collisions are avoided.

Thus, given a road map area which we refer to as the hovering area and a single message information data, we seek to find an information protocol that when applied, all vehicles within the area receive the message, with as low as number of messages exchanged, as possible.

4. Proposed Traffic Adaptive Information Hovering Protocol

The problem of delivering a message to vehicles residing in a specific area, as discussed above, can be intuitively solved through the employment of epidemic routing inside the anchor area (in the subsequent text, hovering area, anchor area and target area are used interchangeably). However, in an area of sparse traffic, the network can be fragmented resulting in partitioned nodes which cannot receive a message broadcasted within other parts of the network. In this case, informed vehicles outside the target area may represent information bridges towards the partitioned nodes. Hence, epidemic routing throughout the entire network can improve reachability at the expense, however, of a significant increase in the number of redundant messages. Therefore, we propose the employment of epidemic routing inside the anchor area and probabilistic flooding outside the anchor area. Probabilistic flooding can considerably limit the number of unnecessary broadcasts.

The operation of our proposed information hovering protocol is the following: first, vehicles are assumed to be aware of their location via a positioning system (i.e., GPS) and of their neighbors through beaconing. The header of an intended message broadcasted by an information source inside the hovering area, must include a time-to-live (TTL) field and an anchor area field. The TTL field controls the expiration of the message and the anchor area field determines location of the hovering area. When the intended packet is transmitted for the first time, all the neighbors of the information source are marked as recipients. Receiving vehicles compare their location with the anchor area field. If a receiving vehicle lies inside the target area, the former periodically inspects its neighbors for uninformed nodes. The scanning frequency is the frequency with which a vehicles verifies that all its neighbors received the intended packet. Further description of the proposed approach is pursued with reference to introduced notation which is summarized in Table 1 for clarity of presentation.

Table 1. Notation used for Protocol Description.

p	Rebroadcast Probability
$f(.)$	$p = f(d)$: rebroadcast probability function
d	distance from the hovering area
r	vehicle transmission range
R	radius of the hovering area

Whenever an informed neighbor is discovered, the vehicle rebroadcasts the intended message and identifies all neighboring vehicles as informed nodes. When, however, the vehicle is outside the anchor area, it rebrodcasts with probability p and does nothing with probability $1 - p$. As mentioned earlier in the paper, the value of p can be dependent on many factors. Here, we calculate p based on a decreasing function, f, of the distance from the target area. The rationale is that the vehicles further away from the anchor area are unlikely to find, and connect, partitioned nodes inside the area. The value of TTL field, which controls the time of expiration of a packet, is locally stored in a receiving vehicle and periodically decremented. When the packet is rebroadcasted, its header TTL field is replaced with the current locally stored TTL value.

The most significant part of the overall design procedure is the determination of the probability function f. Our aim is to determine the appropriate non-increasing function and tune its parameters. In this paper, this is accomplished through a simulation study of different candidate functions and selecting the one that outperforms the others. The overall design procedure is depicted schematically in Figure 1. The first step involves selecting the appropriate rebroadcast probability function out of a set of candidate functions which is the one reporting the highest reachability using the least number of exchanged messages under common simulation settings. The second step involves tuning the parameters of the obtained function by identifying the critical parameter value at which phase transition occurs. This is done for different vehicle densities within the hovering area in order to obtain a critical parameter value vs. density curve. This is repeated for different hovering areas and the resulting curve, used in the proposed solution, is a best fit between the obtained curves. The final step involves a density estimation scheme based on the number of neighbors of each vehicle. The combination of the estimation scheme and the rebroadcast probability curve allows each vehicle to decide independently its rebroadcast probability based on the average number of neighbors and its location.

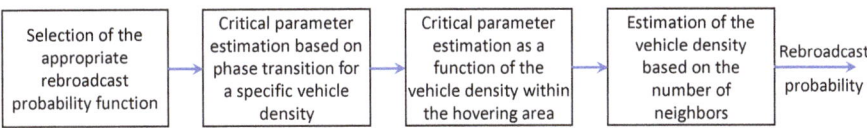

Figure 1. Schematic representation of the design procedure.

The candidate functions chosen are the following: (i) gaussian-like function (Equation (1)); (ii) step function (Equation (2)); (iii) linear function (Equation (3)); (iv) exponential-like function (Equation (4)).

$$\text{Gaussian-like function: } p = e^{-\frac{d^2}{2\sigma^2}} \qquad (1)$$

$$\text{Step function: } p = \begin{cases} 1 & \text{if } d \leq \left(\frac{r}{9}\right) \\ 0.90 & \text{if } \left(\frac{r}{9}\right) < d \leq \left(\frac{r}{4}\right) \\ 0.80 & \text{if } \left(\frac{r}{4}\right) < d \leq \left(\frac{r}{2}\right) \\ 0.70 & \text{if } \left(\frac{r}{2}\right) < d \leq \left(\frac{7r}{9}\right) \\ 0.60 & \text{if } \left(\frac{7r}{9}\right) < d \leq \left(\frac{17r}{18}\right) \\ 0.50 & \text{if } \left(\frac{17r}{18}\right) < d \leq \left(\frac{10r}{9}\right) \\ 0.40 & \text{if } \left(\frac{10r}{9}\right) < d \leq \left(\frac{13r}{9}\right) \\ 0.30 & \text{if } \left(\frac{13r}{9}\right) < d \leq \left(\frac{5r}{3}\right) \\ 0.20 & \text{if } \left(\frac{5r}{3}\right) < d \leq (2r) \\ 0.10 & \text{if } (2r) < d \leq \left(\frac{43r}{18}\right) \\ 0 & \text{if } d > \left(\frac{43r}{18}\right) \end{cases} \qquad (2)$$

$$\text{Linear function: } p = 1 - \frac{1.3d}{3r} \qquad (3)$$

Exponential-like function: $p = e^{-\frac{0.7d}{r}}$ (4)

In Equations (1)–(4), σ represents the standard deviation of the function, d represents the distance between the vehicle and the hovering area, while r indicates the transmission range. These parameters are tuned in order to make candidate probability functions exhibit the similar decreasing behavior shown in Figure 2. It must be noted that decreasing the step size of the step function improves the refinement at the expense of higher implementation overhead. However, the overall patterns reported are not significantly affected.

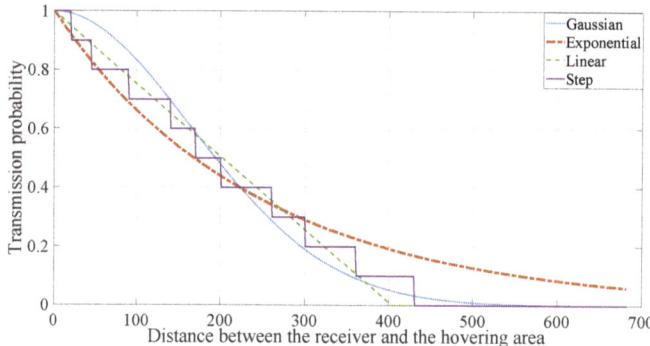

Figure 2. Candidate non-increasing rebroadcast probability functions; the Gaussian, Exponential, Linear, and Step functions.

The reference model used in all simulation experiments, represents a section of the road network in the cities of Bellevue and Redmond in Washington. It includes congested arterial streets and a saturated freeway as shown in Figure 3. The modeled freeway includes lanes for general traffic, HOV/Bus lanes, three closely spaced interchanges, six ramp meters, and two collector-distributor roads. We identify six hovering areas in this road network which are shown in Figure 3, and for this set of experiments we consider area A. We conduct all the simulation experiments using VISSIM [78], a microscopic simulation tool. The aforementioned candidate non-increasing functions are incorporated in the said information hovering protocol producing four corresponding techniques. We evaluate the performance of these schemes with respect to the average reachability and the total number of received packets. The measurements of reachability are conducted for the simulation period after the system enters an equilibrium state.

Furthermore, the results achieved by the probabilistic schemes are compared with those of employing (i) **epidemic routing in the entire network**, that is, $p = 1$ and (ii) **epidemic routing solely inside the anchor area**, that is, $p = 0$. The first technique achieves high reachability at the cost of increased number of the exchanged packets. The second technique on the other hand, minimizes the number of exchanged packets but can achieve low reachability levels in sparse traffic.

Figure 4a depicts the number of received messages, whereas Figure 4b shows the corresponding reachability values. As expected, epidemic routing in the entire network achieves the highest reachability levels and the highest number of messages. Whereas at the bottom of the reachability and the number of messages figures, lie the results of epidemic routing only inside the anchor area and the linearly decreasing probability function. While the rest of the schemes achieve reachability levels comparable to those of overall epidemic routing, they result in smaller numbers of messages. Since the Gaussian-like probability function reports the lowest overhead, it is selected for our intended information hovering protocol.

$$p = \frac{1}{\sigma\sqrt{2\pi}} e^{-\frac{d^2}{2\sigma^2}}.$$ (5)

The second phase of our design procedure, after selecting the proper function, is the tuning of the parameter of this function for the best possible performance. The Gaussian-like function in Equation (5) is obtained by considering a zero mean normal curve with a standard deviation of σ. In order to maintain a rebroadcast probability of 1 at the edges of the target area, the normal curve is multiplied by $\sigma\sqrt{2\pi}$. The value of σ indicates the area where vehicles rebroadcast with a high probability. A scenario with high network connectivity, implies the presence of a large number of rebroadcasting opportunities and thus the area of high rebroadcasting probability must be small to avoid overloading the network. On the other hand, in a network of low connectivity more nodes must rebroadcast with a higher probability to ensure high reachability. This shows that the value of σ must be adaptive to the connectivity of the network inside the anchor area.

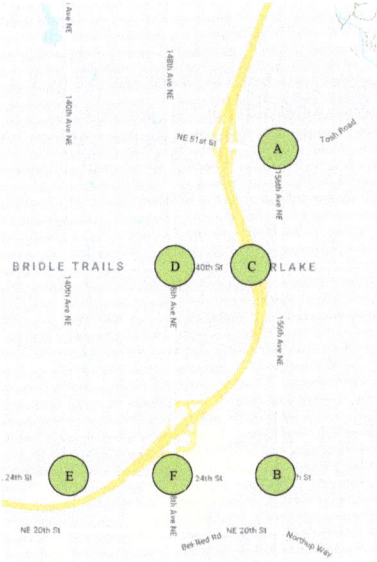

Figure 3. A section of the road network in the cities of Bellevue and Redmond in Washington, used in the simulations reference model. Hovering areas are marked as green circles.

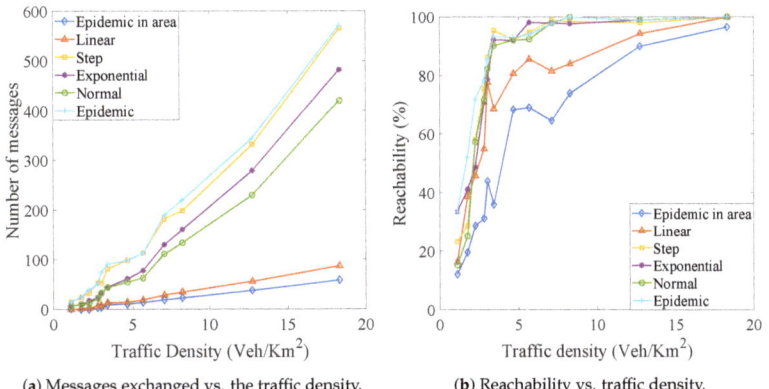

Figure 4. Number of messages exchanged and the Reachability vs. the traffic density for each of the candidate probability functions; epidemic routing inside the area (dark blue), linear (red), step (yellow), exponential (purple), normal (green), and epidemic routing in the entire network (light blue).

Although it is difficult to obtain the connectivity of the network inside the target area through measurements, there are other network parameters which can give an indication of the connectivity. The value of σ can then be adaptively regulated according to the values of these parameters which can be locally available to each vehicle or estimated online. One such parameter is the transmission range of each vehicle. Here, we consider a transmission range of 180 m which is commonly considered for vehicular communications [79]. We leave studying the effect of the transmission range on the performance of the proposed protocol for future publications. The network connectivity is also dependent on the vehicle density. A low (high) density scenario implies low (high) network connectivity. Therefore, high densities inside the hovering area indicates that a small value of σ is sufficient for high reachability results. Through simulations, we define a function mapping density to the corresponding optimal σ. Various vehicle densities are considered. For each vehicle density, σ yielding the best performance is extracted. Despite the fact that the original simulation study is conducted in the mentioned area, the same design procedure is later applied in other hovering areas demonstrating the robustness of the derived function with respect to changing topologies.

In order to characterize the σ yielding the best performance, we fix the traffic density in an area at 6.7 vehicles/Km^2, vary the value of σ between 0 and 405 and observe the reported number of received packets and the reachability level for each σ as depicted in Figure 5a,b, respectively. As expected, larger values of σ result into higher numbers of received packets. However, whereas the reachability-Sigma relationship starts as an increasing function, after a particular value of σ the reachability enters a saturation phase where it attains values close to 95%. This particular value of σ is referred to as the critical value. For any vehicle density there exist a critical value of σ which minimizes the number of packets while maintaining high reachability values. The existence of this critical value is a strong indication of the presence of phase transition phenomena which are typical in the theory of percolation theory and random graphs and are also observed in VANETS in Reference [80]. The same type of behavior is observed in other hovering areas.

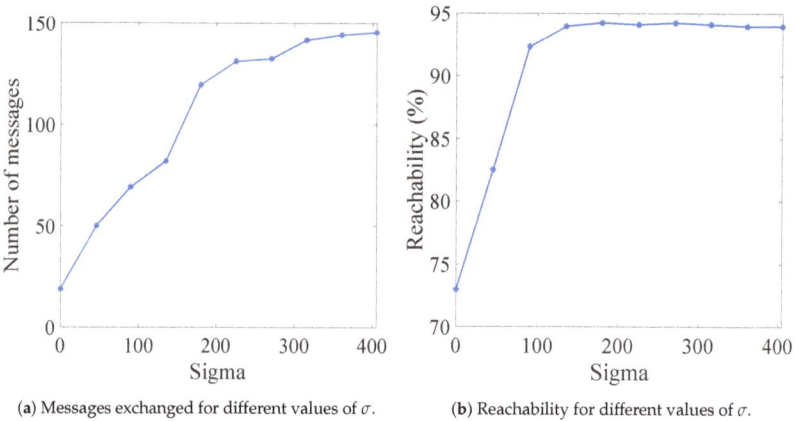

(a) Messages exchanged for different values of σ. (b) Reachability for different values of σ.

Figure 5. Number of messages exchanged and reachability in hovering area A for different values of σ when the vehicle density is set to 6.7 vehicles/Km^2.

In Figure 6a the relationship between reachability and the number of packets in area B is shown for a vehicle density of 15 vehicles/Km^2 which indicates a scenario of a very high density. In Figure 6b, on the other hand, the same relationship is depicted for area C when the vehicle density is 4.9 vehicles/Km^2. The number of exchanged messages, as discussed earlier, can be varied through changing the value of σ. Thus, in the figure, each data point corresponds to a different value of σ (i.e., larger number of packets indicates higher σ values). We consider values of σ in the range 0–405. A value of σ equal to zero implies the employment of epidemic routing in the target area only, whereas a value of $\sigma = \infty$ implies the application of epidemic routing in the entire network. The graphs indicate the existence of a critical value of σ, beyond which the achieved reachability is almost constant. Such critical values of σ are obtained for values of the traffic density in the range 2–16 vehicles/Km^2 in all hovering areas depicted in Figure 3. Figure 7a shows that the relationship between the critical value of σ and the vehicle density is exponentially decreasing for all the considered anchor areas. Hence, the topology of the roadway has nearly no impact on the relationship and thus, a generalized function of the former can be derived. Here, we employ a least square fit between the curves to obtain such a function.

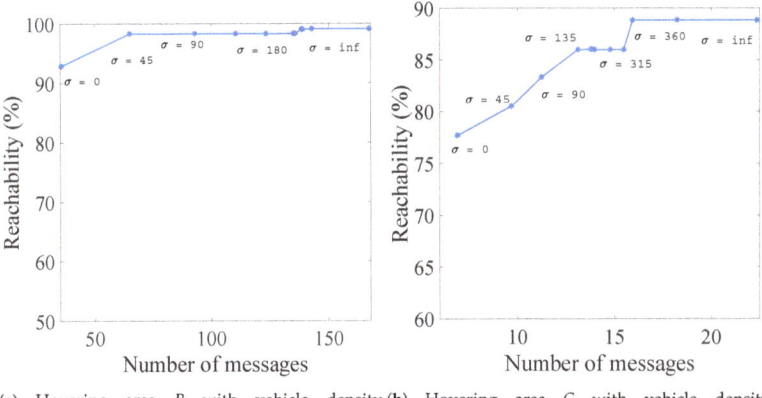

(a) Hovering area B with vehicle density 15 vehicles/Km^2.

(b) Hovering area C with vehicle density 4.9 vehicles/Km^2.

Figure 6. Reachability vs. the number of messages exchanged for different values of σ in areas B and C.

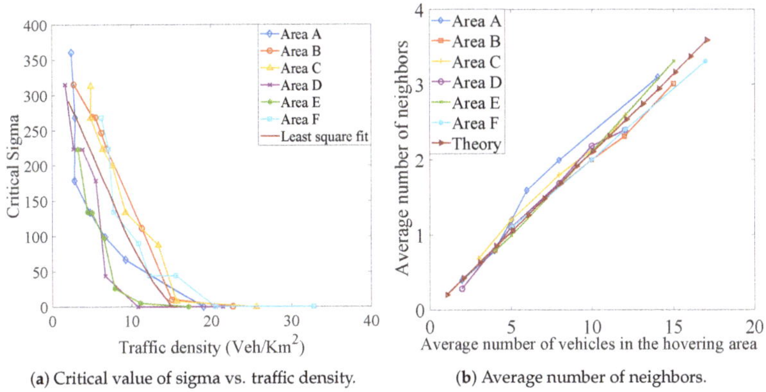

(a) Critical value of sigma vs. traffic density.

(b) Average number of neighbors.

Figure 7. (a) Critical value of sigma vs. traffic density in different hovering areas together with their least squares curve. (b) Average number of neighbors in different hovering areas and average number of vehicles, and comparison with the predicted values using Equation (14).

5. Estimation of the Vehicle Density

Accurate computation of a global information as the vehicle density at the node level is a difficult task, especially in intermittently connected networks. Therefore, for the often partitioned network that we consider, we resort to distributed algorithms in order to estimate the vehicle density online. The vehicle density is then incorporated in determining the sufficient rebroadcast probability outside the anchor area in the proposed information hovering protocol. The proposed vehicle density estimation mechanism relies on the average number of neighbors within the target area. The question is how to relate the vehicular density to the average number of neighbors in a roadway setting. The numerous ways a road network can be constructed renders the derivation of a general relationship between density and the number of neighbors infeasible. Therefore, we consider a simple model of an intersection of two two-way roads where the intersection is at the center of the circular hovering area, as depicted in Figure 8. Significant introduced notation is summarized in Table 2.

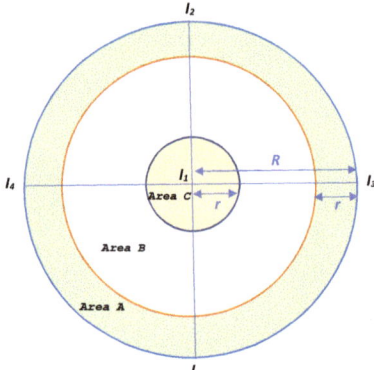

Figure 8. Road topology of the model used to relate analytically the vehicle density vs. the average number of neighbors.

Table 2. Significant Notation for the Density Estimation Scheme.

$n+1$	number of vehicles residing in the hovering area
I_i	roadway edges
n_{i-j}	directional roadway link between I_i and I_j
v_i	vehicle i
R	radius of the anchor area
p_i^h	probability that vehicle i resides in hovering area h
P_{n_k}	probability that v_j is neighbor of v_i given that the latter is in hovering area h
S_t	average number of neighbors
H_t^j	number of neighbors of the jth neighbor

We aim to calculate the expected number of neighbors of each vehicle when the number of vehicles lying in the anchor area is equal to $n+1$. In such case, the vehicle density is expressed by Equation (6), where R denotes the radius of the anchor area.

$$\frac{n+1}{\pi R^2}. \tag{6}$$

At any time instant, we assume that the location of each vehicle is randomly uniformly distributed over all possible locations in the considered roadway, and that two vehicles are considered to be neighbors when they lie within their transmission range. Considering the previous assumption, it follows that the events of any two vehicles being neighbors are independent, and the event of any two vehicles being neighbors is a bernoulli trial with probability of success equal to p. From the independence of the bernoulli trials, the number of neighbors of any vehicle is a binomially distributed random variable whose expected value is np. So, in order to calculate the expected number of neighbors when the number of vehicles in the hovering area is equal to $n+1$, it suffices to calculate the probability p of two vehicles being neighbors. We calculate this probability below.

The basic notations utilized in the subsequent analysis are defined next. Figure 8 depicts the considered hovering area where a pair of two-way straight line roads perpendicularly intersect at point I_1. The one road extends from I_2 to I_5 and the other extends from I_3 to I_4. We consider $n+1$ vehicles within the hovering area. Our objective is to determine the probability of two randomly selected vehicles to be neighbors, that is the distance between then is less than the fixed transmission range r. The h_{i-j} represents the directional roadway link between points I_i and I_j. The direction of the traffic flow is from I_i to I_j. We partition the hovering area in three subareas A, B, C, as shown in Figure 8.

The shaded outer ring bounded from the outside by the circle of radius R and from the inside by the circle of radius $R - r$ represents area A as shown in Figure 8. Area B is the mid ring bounded by the circle of radius $R - r$ and the circle of radius r from the outside and the inside, respectively. Finally, area C covers the most inner circle of radius r. As shown in the figure the circles are all concentric.

The probability of vehicle v_1 residing in area A is denoted by P_1^A. Similarly, we define P_1^B and P_1^C. Given that v_1 lies in area A, the probability that it has v_2 as a neighbor is constant and denoted by Pn_A. Similarly, we define Pn_B and Pn_C. Due to areas A, B, and C constituting a partition of the hovering area, the desired probability p is given by Equation (7).

$$p = P_1^A * Pn_A + P_1^B * Pn_B + P_1^C * Pn_C. \quad (7)$$

Probability P_1^A is given by the ratio between the length of the road sector within area A and the length of the road sector within the anchor area. The lengths of the road in Area A and the road in the anchor area are equal to $8r$ and $8R$, respectively, which yields:

$$P_1^A = \frac{r}{R}. \quad (8)$$

Similarly, we can deduce that:

$$P_1^B = \frac{R - 2r}{R} \quad (9)$$

$$P_1^C = \frac{r}{R}. \quad (10)$$

The probability Pn_A is calculated as follows. First, let v_1 reside on road h_{1-2} in at distance k from the boundary of the area. v_1 and v_2 can be neighbors with a probability of $Pn_{1-2}^A(k)$, where k is in the range of $0 - r$. Therefore, the conditional probability that v_1 is one of v_2's neighbors while residing on h_{1-2} in A, is expressed as $\frac{1}{r}\int_0^r Pn_{1-2}^A(k)dk$. v_2 is declared a neighbor of v_1 in either of the following cases: (a) if it is residing in any side of the road between v_1 and I_2 and (b) if it is residing in any side of the road within a distance r from v_1 in between v_1 and I_1. The probability that the first case is valid is equal to the length of the roadway section between v_1 and I_2 over the total length of the roadway in the hovering area, whereas the probability of the second case being valid due to its geometry is equal to $\frac{1}{4}P_1^A$. So, $Pn_{1-2}^A(k)$ is equal to $\frac{2r}{8R} + \frac{2k}{8R}$ and Pn_{1-2}^A is given by $\frac{1}{r}\int_0^r Pn_{1-2}^A(k)dk = \frac{3r}{8R}$. Similarly we define Pn_{1-j}^A and Pn_{j-1}^A for $j = \{2,3,4,5\}$. Moreover, we define P_{1-j}^A and P_{j-1}^A $j = \{2,3,4,5\}$ to be the probabilities that v_1 lies in road section h_{1-j}, h_{j-1} respectively in area A. It follows that $Pn_A = \sum_{j=2}^{5} Pn_{1-j}^A P_{1-j}^A + Pn_{j-1}^A P_{j-1}^A$. Due to the symmetry of the problem $Pn_{1-j}^A = Pn_{j-1}^A = Pn_{1-2}^A$ $\forall$ $j = \{2,3,4,5\}$ and $P_{1-j}^A = P_{j-1}^A = \frac{1}{8}$ $\forall j = \{2,3,4,5\}$. It follows that:

$$Pn_A = \frac{1}{8} * Pn_{1-2}^A * 8 = \frac{3r}{8R}. \quad (11)$$

In area B, the fact that v_1 is located on road h_{1-2}, implies that v_2 is a neighbor of v_1 if it is less than r apart from v_1 in both directions on h_{1-2} or in road h_{2-1}. In such a case, $Pn_{1-2}^B(k)$ is equal to

$\frac{4r}{8R}$ and Pn_{1-2}^B is expressed by $\frac{1}{R-2r}\int_r^{R-r} Pn_{1-2}^B(k)dk = \frac{4r}{8R}$. Using the same arguments leading to Equation (11), it follows that

$$Pn_B = \frac{1}{8} * Pn_{1-2}^B * 8 = \frac{r}{2R}. \tag{12}$$

Finally, v_1 and v_2 are considered neighbors when v_1 is located in area C on road h_{1-2}, while v_2 is lying at a distcance smaller than r from v_1 on roads h_{1-2} and h_{2-1}. Another scenario where v_1 is a neighbor of v_2 is when the former lies at a distance k from the intersection at I_1 on road h_{1-2} and v_2 resides within a distance $\sqrt{r^2 - k^2}$ from I_1 on roads h_{1-3}, h_{3-1}, h_{1-4}, h_{4-1}. So, given that v_1 is in area C, the probability that the two vehicles are neighbors Pn_{1-2}^C is given by $\frac{4r}{8R} + 4\int_0^r \frac{\sqrt{r^2-k^2}}{8Rr}dk = \frac{r}{2R} + \frac{1}{2Rr}\int_0^r r\sqrt{1-(\frac{k}{r})^2}$. By setting $\frac{k}{r} = \sin\theta$, we get: $Pn_{1-2}^C = \frac{r}{2R} + \frac{1}{2Rr}r\int_0^{\frac{\pi}{2}}\sqrt{1-\sin^2\theta}\, r\cos\theta d\theta = \frac{r}{2R} + \frac{r}{4R}\int_0^{\frac{\pi}{2}} 1+\cos 2\theta d\theta = \frac{r}{2R} + \frac{r\pi}{8R} = \frac{4r+r\pi}{8R}$. Using the same arguments leading to Equation (11), it follows that

$$Pn_C = \frac{1}{8} * Pn_{1-2}^C * 8 = \frac{4r+r\pi}{8R}. \tag{13}$$

Substituting Equations (8)–(13) in (7), Equation (14) is obtained. This implies that the expected number of neighbors $E[X]$, where X denotes the random variable of the number of neighbors of a vehicle in the hovering area, when the vehicle density is equal to $\frac{n+1}{\pi R^2}$ is given by Equation (15):

$$p = \frac{r}{R}\frac{3r}{8R} + \frac{R-2r}{R}\frac{r}{2R} + \frac{r}{R}\frac{4r+r\pi}{8R} = \frac{4Rr + \pi r^2 - r^2}{8R^2} \tag{14}$$

$$E[X] = n\frac{4Rr + \pi r^2 - r^2}{8R^2}. \tag{15}$$

The expressions above are obtained through considering a simple model simple model of a roadway system consisting of two intersecting roads only. Next, we investigate the extent to which our theoretical findings agree with a simulation experiment of a scenario with more relaxed settings. The results show prominent agreement between the theoretical findings and the simulation results despite the simplicity of the theoretical model. We consider Figure 3 and use all the hovering areas (A-F). All considered hovering areas are circular with radius $R = 500$ m. However, each area has a distinct roadway system that is more complex than the topology of the theoretical model, comprising multiple intersecting roads. Our simulation setup assumes a transmission range of $r = 180$ m and several values of the vehicle density in each anchor area. For each vehicle density, simulation results are employed to obtain the average number of neighbors.

The results above are then compared with the function of Equation (14), which becomes $E[X] = 0.21n$ when R and r are assigned the values considered here. In Figure 7b good agreement between the simulation results and our theoretical findings is observed, indicating that Equation (14) may serve the proposed hovering protocol. In order to estimate the average number of neighbors, each vehicle in the hovering area maintains the following state variables:

(i) estimation of the average number of neighbors in the anchor area (S_t);
(ii) number of neighbors n;
(iii) number of neighbors of each of its neighbors (H_t^j for j neighbor);
(iv) estimates of the average number of neighbors of each of its neighbors(S_t^j for the neighbor j).

The information of (ii), (iii) and (iv) are obtained through the exchange of beacon messages whose implementation details are not discussed in this paper. The following exponentially weighted moving average like algorithm is employed to adjust S_t:

$$S_t = a * \frac{(n + \sum_{j=1}^n H_t^j)}{n+1} + (1-a) * \frac{(S_{t-1} + \sum_{j=1}^n S_{t-1}^j)}{n+1} \tag{16}$$

$$S_t = \frac{(S_{t-1} + \sum_{j=1}^{n} S_{t-1}^j)}{n+1}. \tag{17}$$

It is important to mention that this calculation is performed only by the nodes inside the anchor area. Vehicles outside the target area update their information about neighboring nodes through beaconing including information. A vehicle, upon receiving an average neighbor value (S_{t-1}^j) from neighbor j, updates its estimate of the average number of neighbors within the hovering area S_t according to Equation (17). Updates outside the hovering area are essential, as they ensure that most recent updates which are provided by vehicles exiting the hovering area quickly spread in the entire network and in addition improve the robustness of the system with respect to erroneous measurements.

6. Performance Evaluation

In this section, we assess the performance of the proposed protocol through simulation experiments. In these simulations we consider hovering areas, that were not considered in the design process, in the same previously considered road network which includes congested arterial streets and saturated freeway in the cities of Bellevue and Redmond in Washington. The considered areas are indicated in Figure 9 by the letters H and G. The experiments are conducted on VISSIM [78], a microscopic simulation tool. The VISSIM model is tuned priori to generate realistic mobility traces of the vehicles within the considered areas. These traces are then processed by custom made scripts realizing the protocols under consideration for evaluation purposes. We evaluate the performance in terms of the achieved reachability and communication overhead indicated by the number of packets.

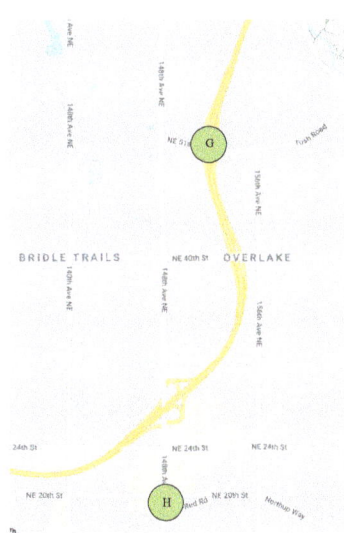

Figure 9. Road topology of the simulations reference model used for protocol evaluation.

6.1. Effectiveness of Adaptive Probabilistic Flooding

We first conduct a comparative study in order to investigate to what extent the proposed scheme achieves its design objectives relative to other schemes which have appeared in literature: (i) epidemic routing in the hovering area only; (ii) epidemic routing in the entire vehicular network; (iii) and the scheme proposed in Reference [28], which considers transmitting the intended message in a finite area outside the target area. The way the size of this area is determined is not specified in Reference [28]. Therefore, we assume the size of the extended area to be the same as the that of the anchor area. This is achieved by setting the radius of the area in which message exchange is allowed, equal to $\sqrt{2}$ times the radius of the hovering area.

Figure 10a,b show the reachability levels and the number of exchanged packets reported by the four considered scheme for different values of the vehicle density, respectively. The low reachability achieved by all methods for small vehicle density values implies the presence of partitioned areas in the network where vehicles cannot receive intended message. Increasing the number of vehicles residing in the anchor area results in an improved reachability and an increased number of exchanged packets. Nonetheless, the proposed protocol outperforms the rest of the considered methods as it reports a small number of exchanged packets while maintaining reachability levels comparable to those achieved by employing epidemic routing within the entire network. Similar results are obtained in area G shown in Figure 11a,b, demonstrating the ability of the proposed scheme to work effectively in various hovering areas with different road topologies and traffic characteristics.

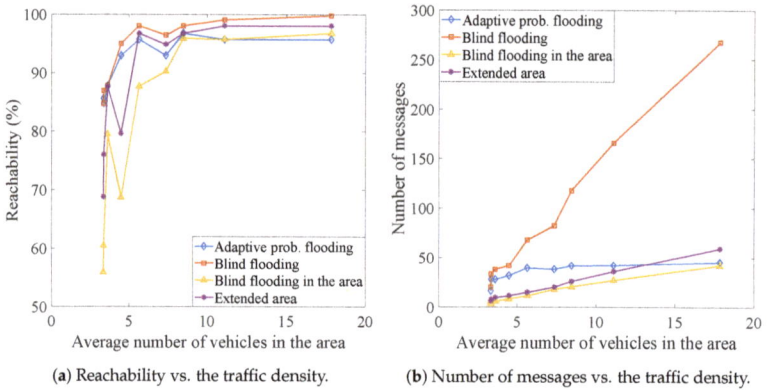

Figure 10. Comparison of the proposed adaptive probabilistic scheme with other hovering schemes in terms of the reachability achieved and the number of messages exchanged in area H for different average number of vehicles in the area.

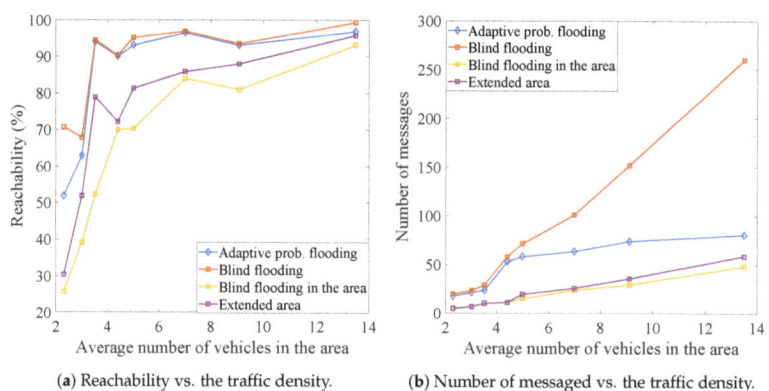

Figure 11. Comparison of the proposed adaptive probabilistic scheme with other hovering schemes in terms of the reachability achieved and the number of messages exchanged in area G for different average number of vehicles in the area.

6.2. Effectiveness of the Estimation Algorithm

In this section we investigate the effect of the estimation algorithm on the performance of the proposed information hovering protocol. The questions we seek to answer relate to how the hovering protocol would perform had the actual vehicle density within the hovering area is known, and to what extent this performance is compromised if estimates which are generated by the estimation algorithm proposed in the previous section replace the actual values. Also, we consider the vehicle density within the hovering area known, and hence the hovering protocol uses this known density value to tune the parameter σ of the rebroadcast probability function according to the function depicted in Figure 7a.

We compare the performance of this protocol to the performance of the proposed information hovering protocol which replaces the unknown density values with estimates generated online. Simulation experiments are conducted in both areas G and H and the results are summarized in the plots of Figure 12, which indicate the number of exchanged messages and the reachability achieved by the two protocols as we increase the vehicle density. The two protocols exhibit similar behavior demonstrating that the estimation algorithm does not compromise the performance of the information hovering which could be achieved, had the vehicle density values be known. Therefore, the estimation algorithm can be considered successful in generating estimates of the unknown vehicle density, which are of acceptable accuracy.

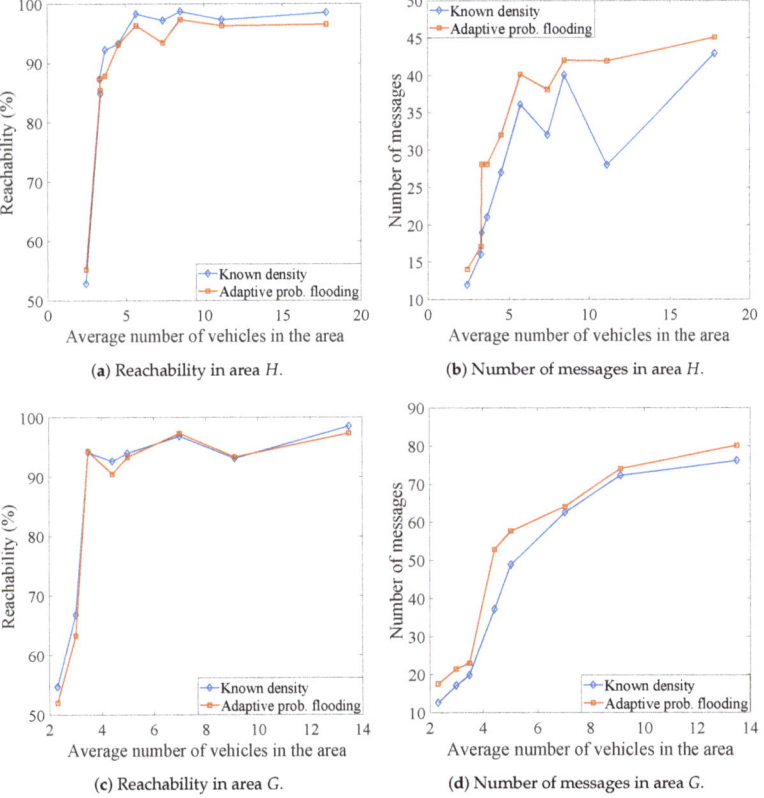

Figure 12. Comparison of the proposed hovering scheme with a protocol which assumes the vehicle density within the hovering area to be known. The comparison is made in terms of the reachability achieved and the number of messages exchanged in areas H and G for different average number of vehicles in the area.

6.3. Comparison with Adaptive Area Regulation

The proposed information hovering protocol aims to solve the problem of low reachability in cases of low traffic density by allowing controlled exchange of messages outside the hovering area. This is achieved by employing adaptive probabilistic flooding; a unique feature of the protocol. The adaptivity mechanism uses estimates of vehicle density, calculated online, to regulate the rebroadcast probability. The information hovering protocol proposed in Reference [28] is based on a different approach which allows message exchange in a restricted area outside the hovering area. As shown earlier, the information hovering protocol proposed in this work outperforms the protocol proposed in Reference [28] when the size of the restricted area is fixed. However, an interesting question is how the two approaches compare if the size of the restricted area had been adaptively regulated based on the vehicle density. The idea of adaptively regulating the area in which message exchange is allowed, based on estimates of the vehicle density, was briefly discussed in Reference [28], however no guidelines have been given on how to design the estimation algorithm and how to use the estimates that it generates to regulate the size of the area. Thus, in order to compare the algorithm proposed in Reference [28] fairly, we make it adaptive by using similar ideas and almost the same design procedure used to design the proposed information hovering protocol.

For the comparative evaluation we consider 12 different implementations of the protocol which applies epidemic routing in an extended circular area encompassing the hovering area itself, with radius R, differing in the size of the extended area. One implementation of the protocol applies epidemic routing in the hovering area only while the radius of the extended area of the rest of the implementation protocol differ by 0.2R. For a particular traffic density in a specific hovering area we compare the performance of the 12 implementations of the protocol and choose the one which exhibits superior performance in terms of high reachability achieved with the smallest possible number of exchanged messages. The radius of the extended area of this implementation is referred to as the critical radius. We repeat the same procedure for different traffic densities and different hovering areas and the resulting critical radius obtained are shown graphically in Figure 13. Similar to the critical sigma curve of Figure 7a, the critical radius curve exhibits an exponentially decreasing pattern as the vehicle density increases. This pattern does not vary significantly in each hovering area indicating that a universal curve may be obtained which is independent of the chosen road topology. Such a curve is extracted by applying a least squares fit between the curves of Figure 13. The least squares fit curve is also shown on the same diagram. Each vehicle estimates the vehicle density within the hovering area utilizing the methodology described in the previous section and uses the least squares curve to calculate the radius of the area in which message exchange is allowed. Then, upon receiving a relevant message each vehicle decides to rebroadcast the message if it lies within the calculated extended area.

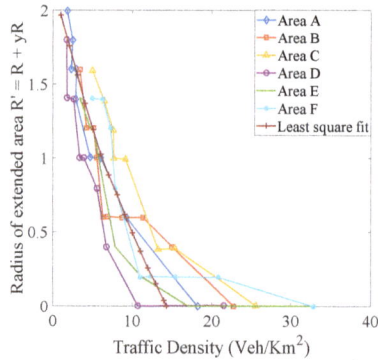

Figure 13. Percentage increase of the critical radius relative to the radius of the hovering area vs. traffic density graphs in different hovering areas, together with their least squares fit used in the adaptive version of Reference [28].

The proposed information hovering protocol and the adaptive version of the protocol proposed in Reference [28] are compared using simulations in areas G and H. The simulation results are shown in Figure 14. We observe almost identical behavior, indicating that any of the two approaches can be used to satisfy the posed design objectives. However, our contribution goes beyond the proposal of a specific information hovering protocol but extends to the introduction of a design procedure which can be used to design a class of density adaptive hovering protocols in other setups.

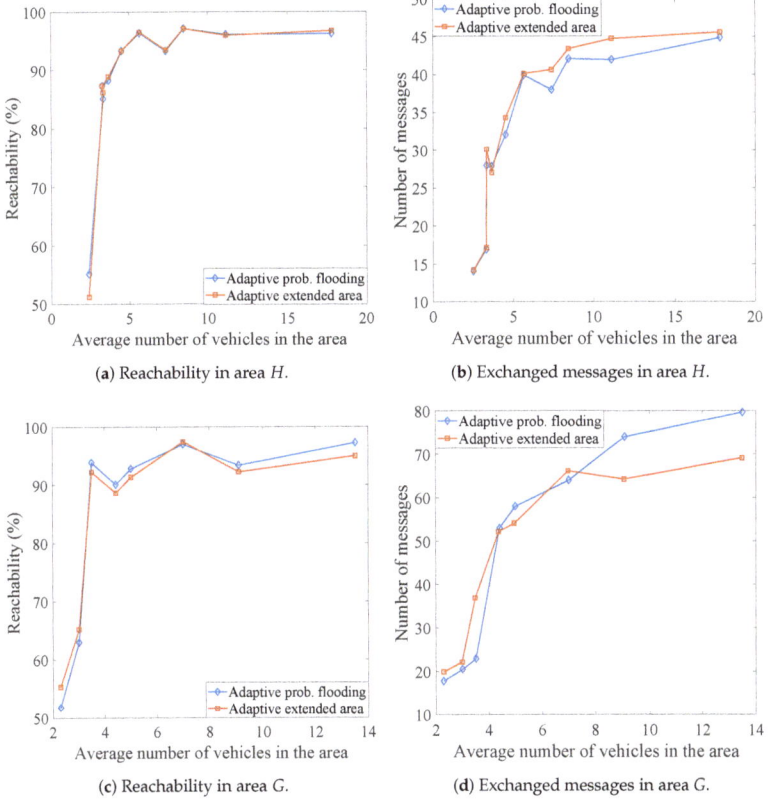

Figure 14. Comparison of the adaptive probabilistic flooding hovering scheme with the adaptive version of the hovering protocol proposed in Reference [28] in terms of the reachability achieved and the number of exchanged messages in areas H and G for different average number of vehicles in the area.

7. Conclusions and Future Work

In this article we address the Information Hovering problem in VANETS, in which useful information needs to be made available to all vehicles within a confined geographical area for a specific time interval. This problem naturally applies not only in VANETS, but in many diverse transportation type applications. The performance of any information hovering protocol is affected by the traffic density of the considered network, especially in cases of low traffic density where partitioned uninformed areas may lead to low reachability. In this work, we propose a novel scheme which overcomes this problem by applying probabilistic flooding outside the hovering area, based on a probability function derived using simulation. Thus, informed vehicles which are outside the area can serve as information bridges towards partitioned uninformed areas, guiding the information back to these areas, hence leading to higher reachability. A unique feature of the proposed protocol is its adaptive mechanism, in which the rebroadcast probability outside the hovering area is adaptively regulated based on estimates of the vehicle density within the hovering area. The estimation of the vehicle density is based on a dynamically derived function. Simulation results show that the proposed protocol is successful in satisfying its design objectives and that it outperforms other candidate hovering protocols. Furthermore, we demonstrate that our contribution extends to the introduction of a design procedure which can be used to design a class of density adaptive hovering protocols, thus going beyond the proposal of a specific information hovering protocol. In order to further support the finding of this study, in the future we aim to use analytical tools from random graph theory.

Author Contributions: Conceptualization, M.L., A.P., A.X., and M.A.; methodology, All; software, A.X. and T.S.; validation, A.X., M.L., and T.S.; Mathematical analysis, A.X., M.A., and M.L.; investigation, All; resources, A.X., A.P., and M.L.; data acquisition, A.X. and T.S.; writing—original draft preparation, All; writing—review and editing, All; supervision, A.P., M.L., and C.M.S.; project administration, A.P. and M.L.; funding acquisition, M.L. and A.P. All authors have read and agreed to the published version of the manuscript.

Funding: This research was partially funded by the EM-VANETS project, CNPq (Conselho Nacional de Desenvolvimento Científico e Tecnológico) grant 303933/2017-8, CAPES (Coordenação de Aperfeiçoamento de Pessoal de Nível Superior), FAPEMIG (Fundação de Amparo à Pesquisa do Estado de Minas Gerais) grant APQ-02145-18, and the Federal University of São João del-Rei (UFSJ).

Conflicts of Interest: The authors declare no conflict of interest.

Abbreviations

The following abbreviations are used in this manuscript:

DTN	Delay Tolerant Networks
MFC	Message Frequency Control
RTT	Round Trip Times
RSUs	Roadside Units
VANETs	Vehicular Adhoc Networks
TPC	Transmit Power Control
VDTNs	vehicular Delay Tolerant Networks
V2V	Vehicle-to-Vehicle
V2I	Vehicle-to-Infrastructure

References

1. Silva, C.M.; Masini, B.M.; Ferrari, G.; Thibault, I. A Survey on Infrastructure-Based Vehicular Networks. *Mob. Inf. Syst.* **2017**, *2017*, 28–56. [CrossRef]
2. Swain, P.; Christophorou, C.; Bhattacharjee, U.; Silva, Cristiano, M.; Pitsillides, A. Selection of UE-based Virtual Small Cell Base Stations using Affinity Propagation Clustering. In Proceedings of the 2018 14th International Wireless Communications Mobile Computing Conference (IWCMC), Limassol, Cyprus, 25–29 June 2018; pp. 1104–1109. [CrossRef]
3. Silva, C.M.; Meira, W., Jr. Managing Infrastructure-Based Vehicular Networks. In Proceedings of the 2015 16th IEEE International Conference on Mobile Data Management (MDM), Pittsburgh, PA, USA, 15–18 June 2015; Volume 2, pp. 19–22. [CrossRef]
4. Eriksson, J.; Girod, L.; Hull, B.; Newton, R.; Madden, S.; Balakrishnan, H. The pothole patrol: using a mobile sensor network for road surface monitoring. In Proceedings of the 6th International Conference on Mobile Systems, Applications, and Services, ACM MobiSys, Breckenridge, CO, USA, 17–20 June 2008.
5. Mukhtar, A.; Xia, L.; Tang, T.B. Vehicle Detection Techniques for Collision Avoidance Systems: A Review. *IEEE Trans. Intell. Transp. Syst.* **2015**, *16*, 2318–2338. [CrossRef]
6. Silva, C.M.; Aquino, A.L.L.; Meira, W., Jr. Smart Traffic Light for Low Traffic Conditions. *Mob. Netw. Appl.* **2015**, *20*, 285–293. [CrossRef]

7. Johnson, D.A.; Trivedi, M.M. Driving style recognition using a smartphone as a sensor platform. In Proceedings of the 2011 14th International IEEE Conference on Intelligent Transportation Systems (ITSC), Washington, DC, USA, 5–7 October 2011; pp. 1609–1615.
8. Oliveira, T.R.; Silva, C.M.; Macedo, D.F.; Nogueira, J.M.S. SNVC: Social networks for vehicular certification. *Comput. Netw.* **2016**, *111*, 129–140. [CrossRef]
9. Silva, C.M.; Silva, F.A.; Sarubbi, J.F.; Oliveira, T.R.; Meira, W., Jr.; Nogueira, J.M.S. Designing mobile content delivery networks for the Internet of vehicles. *Veh. Commun.* **2017**, *8*, 45–55. [CrossRef]
10. Khelifi, H.; Luo, S.; Nour, B.; Moungla, H.; Faheem, Y.; Hussain, R.; Ksentini, A. Named data networking in vehicular ad hoc networks: State-of-the-art and challenges. *IEEE Commun. Surv. Tutor.* **2019**, *22*, 320–351. [CrossRef]
11. Sharma, A.; Awasthi, L.K. A Comparative Survey on Information Dissemination in Heterogeneous Vehicular Communication Networks. In Proceedings of the 2018 First International Conference on Secure Cyber Computing and Communication (ICSCCC), Jalandhar, India, 15–17 December 2018; pp. 556–560.
12. Toor, Y.; Muhlethaler, P.; Laouiti, A. Vehicle ad hoc networks: Applications and related technical issues. *IEEE Commun. Surv. Tutor.* **2008**, *10*, 74–88. [CrossRef]
13. Zeadally, S.; Hunt, R.; Chen, Y.S.; Irwin, A.; Hassan, A. Vehicular ad hoc networks (VANETS): status, results, and challenges. *Telecommun. Syst.* **2012**, *50*, 217–241. [CrossRef]
14. Silva, C.M.; Silva, L.D.; Santos, L.A.L.; Sarubbi, J.F.M.; Pitsillides, A. Broadening Understanding on Managing the Communication Infrastructure in Vehicular Networks: Customizing the Coverage Using the Delta Network. *Future Internet* **2018**, *11*, 1. [CrossRef]
15. Silva, C.M.; Meira, W., Jr. Design of roadside communication infrastructure with QoS guarantees. In Proceedings of the 2015 IEEE Symposium on Computers and Communication (ISCC), Larnaca, Cyprus, 6–9 July 2015; pp. 439–444. [CrossRef]
16. Silva, C.M.; Guidoni, D.L.; Souza, F.S.H.; Pitangui, C.G.; Sarubbi, J.F.M.; Pitsillides, A. Gamma Deployment: Designing the Communication Infrastructure in Vehicular Networks Assuring Guarantees on the V2I Inter-Contact Time. In Proceedings of the 2016 IEEE 13th International Conference on Mobile Ad Hoc and Sensor Systems (MASS), Brasilia, Brazil, 10–13 October 2016; pp. 263–271. [CrossRef]
17. Villalba, A.; Konstantas, D. Towards hovering information. In *Smart Sensing and Context*; Springer: Berlin/Heidelberg, Germany, 2006; pp. 250–254.
18. Castro, A.A.V.; Serugendo, G.D.M.; Konstantas, D. Hovering information–self-organizing information that finds its own storage. In *Autonomic Communication*; Springer: Berlin/Heidelberg, Germany, 2009; pp. 111–145.
19. Di, G.; Serugendo, M. Dependable requirements for hovering information. In *Supplemental Volume—The 37th Annual IEEE/IFIP International Conference on Dependable Systems and Networks (DSN'07)*; IEEE Computer Society Press: Edinburgh, UK, 2007.
20. Shoaib, M.; Song, W.C. Hovering information based vanet applications. In *International Conference on Networked Digital Technologies*; Springer: Berlin/Heidelberg, Germany, 2012; pp. 551–564.
21. Nikolovski, T. Design and Performance Evaluation of Data Dissemination and Hovering Information Protocols for Vehicular Ad Hoc Networks (VANETs). Master's Thesis, University of Ontario Institute of Technology, Oshawa, ON, Canada, 2017.
22. Ciocan, M.; Dobre, C.; Cristea, V.; Mavromoustakis, C.X.; Mastorakis, G. Analysis of vehicular storage and dissemination services based on floating content. In *International Conference on Mobile Networks and Management*; Springer: Berlin/Heidelberg, Germany, 2014; pp. 387–400.
23. Nikolovski, T.; Pazzi, R.W. A Lightweight and Efficient Approach (LEA) for Hovering Information protocols. In Proceedings of the 6th ACM Symposium on Development and Analysis of Intelligent Vehicular Networks and Applications, Valletta, Malta, 13–17 November 2016; pp. 31–38.
24. Xeros, A.; Lestas, M.; Andreou, M.; Pitsillides, A. Adaptive probabilistic flooding for Information Hovering in VANETs. In Proceedings of the Vehicular Networking Conference (VNC), Jersey City, NJ, USA, 13–15 December 2010; pp. 239–246.
25. Ott, J.; Hyytiä, E.; Lassila, P.; Vaegs, T.; Kangasharju, J. Floating content: Information sharing in urban areas. In Proceedings of the 2011 IEEE International Conference on Pervasive Computing and Communications (PerCom), Kyoto, Japan, 11–15 March 2011; pp. 136–146.

26. Konstantas, D.; Villalba, A. Hovering Information: A paradigm for sharing location-bound information. In *Proceedings of the European Conference on Smart Sensing and Context*; Springer: Berlin/Heidelberg, Germany, 2006.
27. Liaskos, C.; Xeros, A.; Papadimitriou, G.I.; Lestas, M.; Pitsillides, A. Balancing wireless data broadcasting and information hovering for efficient information dissemination. *IEEE Trans. Broadcast.* **2012**, *58*, 66–76. [CrossRef]
28. Hermann, S.D.; Michl, C.; Wolisz, A. Time-stable geocast in intermittently connected ieee 802.11 manets. In Proceedings of the 2007 IEEE 66th Vehicular Technology Conference, Baltimore, MD, USA, 30 September–3 October 2007; pp. 1922–1926.
29. Bhaskar, K.; Stephen, B.W.; Ramon, B. Phase Transition Phenomena in Wireless Ad-Hoc Networks. In Proceedings of the GLOBECOM, San Antonio, TX, USA, 25–29 November 2001.
30. Xeros, A.; Lestas, M.; Andreou, M.; Pitsillides, A.; Ioannou, P. Information hovering in vehicular ad-hoc networks. In Proceedings of the 2009 IEEE GLOBECOM Workshops, Honolulu, HI, USA, 30 November–4 December 2009; pp. 1–6.
31. Tseng, Y.C.; Ni, S.Y.; Chen, Y.S.; Sheu, J.P. The broadcast storm problem in a mobile ad hoc network. *Wirel. Netw.* **2002**, *8*, 153–167. [CrossRef]
32. Williams, B.; Camp, T. Comparison of broadcasting techniques for mobile ad hoc networks. In *Proceedings of the 3rd ACM International Symposium on Mobile Ad Hoc Networking & Computing*; ACM: New York, NY, USA, 2002; pp. 194–205.
33. Silva, C.M.; Meira, W., Jr. An architecture integrating stationary and mobile roadside units for providing communication on Intelligent Transportation Systems. In Proceedings of the NOMS 2016—2016 IEEE/IFIP Network Operations and Management Symposium, Istanbul, Turkey, 25–29 April 2016; pp. 358–365. [CrossRef]
34. Boussoufa-Lahlah, S.; Semchedine, F.; Bouallouche-Medjkoune, L. Geographic routing protocols for Vehicular Ad hoc NETworks (VANETs): A survey. *Veh. Commun.* **2018**, *11*, 20–31. [CrossRef]
35. Tian, D.; Liu, C.; Duan, X.; Sheng, Z.; Ni, Q.; Chen, M.; Leung, V.C. A distributed position-based protocol for emergency messages broadcasting in vehicular ad hoc networks. *IEEE Internet Things J.* **2018**, *5*, 1218–1227. [CrossRef]
36. Sung, Y.; Lee, M. A road layout based broadcast mechanism for urban vehicular ad hoc networks. *Wirel. Commun. Mob. Comput.* **2018**, *2018*, 1565363. [CrossRef]
37. Maia, G.; Villas, L.A.; Viana, A.C.; Aquino, A.L.; Boukerche, A.; Loureiro, A.A. A rate control video dissemination solution for extremely dynamic vehicular ad hoc networks. *Perform. Eval.* **2015**, *87*, 3–18. [CrossRef]
38. Oliveira, R.; Montez, C.; Boukerche, A.; Wangham, M.S. Reliable data dissemination protocol for VANET traffic safety applications. *Ad Hoc Netw.* **2017**, *63*, 30–44. [CrossRef]
39. Rayeni, M.S.; Hafid, A.; Sahu, P.K. Dynamic spatial partition density-based emergency message dissemination in VANETs. *Veh. Commun.* **2015**, *2*, 208–222. [CrossRef]
40. Woon, W.; Yeung, K.L. Self-pruning broadcasting for mobile ad hoc networks. In Proceedings of the Global Telecommunications Conference, GLOBECOM 2009, Honolulu, HI, USA, 30 November–4 December 2009; pp. 1–6.
41. Peng, W.; Lu, X.C. On the reduction of broadcast redundancy in mobile ad hoc networks. In *Proceedings of the 1st ACM International Symposium on Mobile ad Hoc Networking & Computing*; IEEE Press: Piscataway, NJ, USA, 2000; pp. 129–130.
42. Bakhouya, M.; Gaber, J.; Lorenz, P. An adaptive approach for information dissemination in vehicular ad hoc networks. *J. Netw. Comput. Appl.* **2011**, *34*, 1971–1978. [CrossRef]
43. Yassein, M.B.; Nimer, S.F.; Al-Dubai, A.Y. A new dynamic counter-based broadcasting scheme for mobile ad hoc networks. *Simul. Model. Pract. Theory* **2011**, *19*, 553–563. [CrossRef]
44. Reina, D.; Toral, S.; Johnson, P.; Barrero, F. A survey on probabilistic broadcast schemes for wireless ad hoc networks. *Ad Hoc Netw.* **2015**, *25*, 263–292. [CrossRef]
45. Liu, W.; Nakauchi, K.; Shoji, Y. A neighbor-based probabilistic broadcast protocol for data dissemination in mobile iot networks. *IEEE Access* **2018**, *6*, 12260–12268. [CrossRef]
46. Cunha, F.; Boukerche, A.; Villas, L.; Viana, A.; Loureiro, A. Data Communication in VANETs: A Survey, Challenges and Applications. *Ad Hoc Netw.* **2016**, *44*, 90–103. [CrossRef]

47. Kaiwartya, O.; Kumar, S. Guaranteed geocast routing protocol for vehicular adhoc networks in highway traffic environment. *Wirel. Pers. Commun.* **2015**, *83*, 2657–2682. [CrossRef]
48. Kaiwartya, O.; Kumar, S. Enhanced caching for geocast routing in vehicular Ad Hoc network. In *Intelligent Computing, Networking, and Informatics*; Springer: Berlin/Heidelberg, Germany, 2014; pp. 213–220.
49. Allani, S.; Yeferny, T.; Chbeir, R. A scalable data dissemination protocol based on vehicles trajectories analysis. *Ad Hoc Netw.* **2018**, *71*, 31–44. [CrossRef]
50. Ko, Y.B.; Vaidya, N.H. Geocasting in mobile ad hoc networks: Location-based multicast algorithms. In Proceedings of the WMCSA'99. Second IEEE Workshop on Mobile Computing Systems and Applications, New Orleans, LA, USA, 25–26 February 1999; pp. 101–110.
51. Allal, S.; Boudjit, S. Geocast routing protocols for vanets: Survey and guidelines. In Proceedings of the 2012 Sixth International Conference on Innovative Mobile and Internet Services in Ubiquitous Computing (IMIS), Palermo, Italy, 4–6 July 2012; pp. 323–328.
52. Maihöfer, C.; Leinmüller, T.; Schoch, E. Abiding geocast: Time–stable geocast for ad hoc networks. In Proceedings of the 2nd ACM International Workshop on Vehicular ad Hoc Networks, Cologne, Germany, 2 September 2005; pp. 20–29.
53. Yu, Q.; Heijenk, G. Abiding geocast for warning message dissemination in vehicular ad hoc networks. In Proceedings of the IEEE International Conference on Communications Workshops, 2008. ICC Workshops' 08, Beijing, China, 19–23 May 2008; pp. 400–404.
54. Soares, V.N.; Rodrigues, J.J.; Farahmand, F. GeoSpray: A geographic routing protocol for vehicular delay-tolerant networks. *Inf. Fusion* **2014**, *15*, 102–113. [CrossRef]
55. Pereira, P.R.; Casaca, A.; Rodrigues, J.J.; Soares, V.N.; Triay, J.; Cervelló-Pastor, C. From delay-tolerant networks to vehicular delay-tolerant networks. *IEEE Commun. Surv. Tutor.* **2012**, *14*, 1166–1182. [CrossRef]
56. Kloiber, B.; Strang, T.; Spijker, H.; Heijenk, G. Improving information dissemination in sparse vehicular networks by adding satellite communication. In Proceedings of the 2012 IEEE Intelligent Vehicles Symposium (IV), Madrid, Spain, 3–7 June 2012; pp. 611–617.
57. Mittag, J.; Thomas, F.; Härri, J.; Hartenstein, H. A comparison of single-and multi-hop beaconing in VANETs. In *Proceedings of the Sixth ACM International Workshop on Vehicular Internetworking*; ACM: New York, NY, USA, 2009; pp. 69–78.
58. Shah, S.A.A.; Ahmed, E.; Xia, F.; Karim, A.; Shiraz, M.; Noor, R.M. Adaptive beaconing approaches for vehicular ad hoc networks: A survey. *IEEE Syst. J.* **2016**, *12*, 1263–1277. [CrossRef]
59. Boukerche, A.; Rezende, C.; Pazzi, R.W. Improving neighbor localization in vehicular ad hoc networks to avoid overhead from periodic messages. In Proceedings of the Global Telecommunications Conference, GLOBECOM 2009, Honolulu, HI, USA, 30 November–4 December 2009; pp. 1–6.
60. Schmidt, R.K.; Leinmuller, T.; Schoch, E.; Kargl, F.; Schafer, G. Exploration of adaptive beaconing for efficient intervehicle safety communication. *IEEE Netw.* **2010**, *24*, 14–19. [CrossRef]
61. Tielert, T.; Jiang, D.; Chen, Q.; Delgrossi, L.; Hartenstein, H. Design methodology and evaluation of rate adaptation based congestion control for vehicle safety communications. In Proceedings of the 2011 IEEE Vehicular Networking Conference (VNC), Amsterdam, The Netherlands, 14–16 November 2011; pp. 116–123.
62. Kloiber, B.; Harri, J.; Strang, T. Dice the TX power—Improving Awareness Quality in VANETs by random transmit power selection. In Proceedings of the 2012 IEEE Vehicular Networking Conference (VNC), Seoul, Korea, 14–16 November 2012; pp. 56–63.
63. Torrent-Moreno, M.; Mittag, J.; Santi, P.; Hartenstein, H. Vehicle-to-vehicle communication: fair transmit power control for safety-critical information. *IEEE Trans. Veh. Technol.* **2009**, *58*, 3684–3703. [CrossRef]
64. Burleigh, S.; Hooke, A.; Torgerson, L.; Fall, K.; Cerf, V.; Durst, B.; Scott, K.; Weiss, H. Delay-tolerant networking: an approach to interplanetary internet. *IEEE Commun. Mag.* **2003**, *41*, 128–136. [CrossRef]
65. Fall, K.; Scott, K.L.; Burleigh, S.C.; Torgerson, L.; Hooke, A.J.; Weiss, H.S.; Durst, R.C.; Cerf, V. Delay-Tolerant Networking Architecture RFC 4838, April 2007. Available online: https://tools.ietf.org/html/rfc4838 (accessed on 10 June 2020).
66. Scott, K.; Burleigh, S. Bundle Protocol Specification, IETF RFC 5050, Experimental, November 2007. Available online: https://tools.ietf.org/html/rfc5050 (accessed on 10 June 2020).
67. Soares, V.N.; Farahmand, F.; Rodrigues, J.J. A layered architecture for vehicular delay-tolerant networks. In Proceedings of the IEEE Symposium on Computers and Communications, ISCC 2009, Sousse, Tunisia, 5–8 July 2009; pp. 122–127.

68. Benamar, N.; Singh, K.D.; Benamar, M.; El Ouadghiri, D.; Bonnin, J.M. Routing protocols in vehicular delay tolerant networks: A comprehensive survey. *Comput. Commun.* **2014**, *48*, 141–158. [CrossRef]
69. Mangrulkar, R.; Atique, M. Routing protocol for delay tolerant network: A survey and comparison. In Proceedings of the 2010 IEEE International Conference on Communication Control and Computing Technologies (ICCCCT), Ramanathapuram, India, 7–9 October 2010; pp. 210–215.
70. Spyropoulos, T.; Psounis, K.; Raghavendra, C.S. Spray and wait: An efficient routing scheme for intermittently connected mobile networks. In *Proceedings of the 2005 ACM SIGCOMM Workshop on Delay-Tolerant Networking*; ACM: New York, NY, USA, 2005; pp. 252–259.
71. Jain, S.; Fall, K.; Patra, R. Routing in a delay tolerant network. In Proceedings of the 2004 Conference on Applications, Technologies, Architectures, and Protocols for Computer Communications, Portland, OR, USA, 30 August–3 September 2004; Volume 34.
72. Leontiadis, I.; Mascolo, C. GeOpps: Geographical opportunistic routing for vehicular networks. In Proceedings of the 2007 IEEE International Symposium on a World of Wireless, Mobile and Multimedia Networks, Helsinki, Finland, 18–21 June 2007; pp. 1–6.
73. Zhu, X.; Chen, C.; Pan, Z.; Guan, X. Fast and Reliable Warning Message Dissemination Mechanism in Urban Expressway VANETs. In Proceedings of the IEEE/CIC International Conference on Communications in China, Chengdu, China, 27–29 July 2016; pp. 1–6.
74. Liu, Y.; Chen, C.; Guan, X. A hovering-based warning information dissemination approach in highway entrances. In Proceedings of the 2014 IEEE International Conference on Communications (ICC), Sydney, Australia, 16–20 June 2014; pp. 2719–2724.
75. Shuai, X. Neighbor Trigger for Hovering Information. *J. Phys. Conf. Ser.* **2018**, *1060*, 012039. [CrossRef]
76. Amjad, Z.; Song, W.C.; Ahn, K.J. Context-Aware Routing for hovering information in Vehicular Ad-Hoc Networks. In Proceedings of the 2016 18th Asia-Pacific Network Operations and Management Symposium (APNOMS), Kanazawa, Japan, 5–7 October 2016; pp. 1–6.
77. Song, W.C.; Ur Rehman, S.; Awan, M.B. Road Aware Information Sharing in VANETs. *KSII Trans. Internet Inf. Syst.* **2015**, *9*. [CrossRef]
78. PTV Traffic Mobility Logistics. Available online: http://www.ptvamerica.com/vissim.htm (accessed on 10 June 2020).
79. Silva, C.M.; Meira, W.; Sarubbi, J.F.M. Non-Intrusive Planning the Roadside Infrastructure for Vehicular Networks. *IEEE Trans. Intell. Transp. Syst.* **2016**, *17*, 938–947. [CrossRef]
80. Mylonas, Y.; Lestas, M.; Pitsillides, A.; Ioannou, P.; Papadopoulou, V. Speed Adaptive Probabilistic Flooding for Vehicular Ad Hoc Networks. *IEEE Trans. Veh. Technol.* **2015**, *64*, 1973–1990. [CrossRef]

© 2020 by the authors. Licensee MDPI, Basel, Switzerland. This article is an open access article distributed under the terms and conditions of the Creative Commons Attribution (CC BY) license (http://creativecommons.org/licenses/by/4.0/).

Article

Diagnosing Automotive Damper Defects Using Convolutional Neural Networks and Electronic Stability Control Sensor Signals

Thomas Zehelein *, Thomas Hemmert-Pottmann and Markus Lienkamp

Institute of Automotive Technology, Technical University of Munich, Boltzmannstr. 15, 85748 Garching, Germany; thomas.hemmert-pottmann@tum.de (T.-H.P.); Lienkamp@ftm.mw.tum.de (M.L.)
* Correspondence: thomas.zehelein@tum.de

Received: 11 December 2019; Accepted: 10 January 2020; Published: 16 January 2020

Abstract: Chassis system components such as dampers have a significant impact on vehicle stability, driving safety, and driving comfort. Therefore, monitoring and diagnosing the defects of these components is necessary. Currently, this task is based on the driver's perception of component defects in series production vehicles, even though model-based approaches in the literature exist. As we observe an increased availability of data in modern vehicles and advances in the field of deep learning, this paper deals with the analysis of the performance of Convolutional Neural Networks (CNN) for the diagnosis of automotive damper defects. To ensure a broad applicability of the generated diagnosis system, only signals of a classic Electronic Stability Control (ESC) system, such as wheel speeds, longitudinal and lateral vehicle acceleration, and yaw rate, were used. A structured analysis of data pre-processing and CNN configuration parameters were investigated in terms of the defect detection result. The results show that simple Fast Fourier Transformation (FFT) pre-processing and configuration parameters resulting in small networks are sufficient for a high defect detection rate.

Keywords: automotive; damper; convolutional neural networks; fault detection; diagnosis; machine learning; deep learning

---

1. Introduction

Ensuring driving safety and driving comfort when operating vehicles requires their health state to be properly monitored. This is especially critical for chassis system components such as dampers. Currently, in addition to the driver's perception, there is only periodic human inspection for monitoring the vehicle's chassis system health state. However, this is error-prone, expensive, and implies periods of unmonitored driving between inspections. Furthermore, autonomous driving implies that the driver is not needed as a monitoring instance, either for the actual driving task or for monitoring the vehicle's health state. Therefore, an automated system for this task is necessary.

Approaches in the field of Fault Detection and Isolation (FDI) can be categorized as reliability-based, model-based, signal-based, and statistical-based FDI [1]. Existing approaches in monitoring the health of the chassis system of a vehicle are often model-based [2–4] or signal-based [5,6]. However, to the authors' knowledge, there is no such approach applied in a series production vehicle. Possible reasons are that either additional sensors that are not part of a vehicle's standard sensor set (e.g., vertical acceleration sensors) are necessary or measurements at a test-bench are required. One problem in automotive damper defect diagnosis during actual driving is robustness with regard to the vehicle's configuration, e.g., changing tire characteristics, mass variations in the vehicle, or varying road excitation. Even though driving data incorporate these different vehicle configurations and are generated while driving, the named approaches cannot benefit directly from more data. The named approaches need to get fine-tuned, which is time-consuming to match the data.

Data-driven approaches are based on measurement data that are available from a process [7]. Combining a signal-based with a data-driven approach leads to machine-learning algorithms. Robustness is therefore automatically incorporated when the supplied training data cover variations of circumstances regarding different vehicle configurations and usage scenarios. A machine-learning approach for automotive damper health monitoring using a Support Vector Machine (SVM) for classifying signal features is presented in [8]. One downside of this approach is the fact that features that can distinguish between different health states are required. Engineering representative features is therefore necessary, which is also time-consuming and requires system knowledge.

Deep learning classification architectures are able to learn features directly from the input data. The parameters of the network are adjusted with respect to minimizing a cost function that accounts for the classification result. Therefore, the overall algorithm is trained with respect to distinguishing between different states of the data. However, adding their increased complexity compared to traditional machine learning algorithms (e.g., SVM) makes sense only if these simple algorithms are lacking performance. Based on the classification results in [8,9], applying deep learning algorithms for damper defect detection should be investigated.

Convolutional Neural Networks (CNN) are stated to be able to deal with multidimensional data as well as having a good local feature extraction [10]. An overview of the different applications of CNN regarding machine health monitoring is given in [11]. CNN have emerged into a broad variety of fields, such as predictive maintenance [12,13] and medical [14,15], or mechanical diagnosis [16–23]. The latter has been dominated by model-driven approaches for decades and more recently, data-driven approaches based on feature engineering. In the past couple of years, researchers have investigated and successfully employed CNN's feature learning capabilities to specifically diagnose rotating mechanical applications such as bearings. However, to the best of the authors' knowledge, there is no application of CNN for the diagnosis of automotive suspension components such as dampers. It is therefore an open question whether CNN are equally suited for the diagnosis of automotive dampers using only Electronic Stability Control (ESC) system sensors and normal driving data. There might be similarities due to sensor signals coming from rotating wheels. However, there are big differences from industrial bearing applications to automotive applications because there is a stochastic excitation of the vehicle caused by the road profile as well as a high variability of circumstances of vehicle usage (e.g., weather conditions or parameter variations such as mass).

This paper investigates the suitability of CNN for the diagnosis of defective automotive dampers. The current state of the art is analyzed in Section 2 with regard to pre-processing methods and network architectures. Section 3 screens various pre-processing methods. Afterwards, experiments with different parameters regarding the size of the receptive field, the size of the pooling layer as well as the network depth of the CNN architecture are conducted and the resulting kernel weights of the trained networks are analyzed. Section 4 evaluates the robustness of the generated diagnosis systems with regard to variations of the vehicle setup. The paper closes with a discussion and summary.

2. State of The Art

This section analyzes the state of the art regarding pre-processing methods and network configurations for diagnosis applications. Hereby, many approaches deal with bearing or gearbox applications. Since CNN emerged from computer vision with two-dimensional input data such as pictures, many researchers transform their data into images. But also one-dimensional data (such as time series data or Fast Fourier Transformation (FFT) data) are used as input data to CNN.

Xia et al. [24] have classified the Case Western Reserve University (CWRU) bearing data set of [25] without any pre-processing. Acceleration sensor signals are used directly as input data to a CNN consisting of two convolutional and sub-sampling layers followed by a fully connected layer. Eren et al. [26] have also proposed no pre-processing and process the CWRU bearing dataset. A CNN consisting of three convolutional and two sub-sampling layers followed by a fully connected layer is used. Zilong and Wei [27] have also performed no pre-processing but propose a CNN architecture that consists of multi-scale convolutional layers. Those layers incorporate the thought of "inception" modules from [28] of extracting features with convolutional kernels of different sizes in parallel. Even though it is a rather deep network architecture, the number of trainable parameters of 52,000 is still in a low range. Zhang et al. [18] have designed a CNN to operate on the noisy raw data. They claim that using Dropout, a regularization technique in the first convolutional layer helps to improve anti-noise abilities and suggest a twelve layer-deep network. Additionally, a very small batch size during training is used as well as an ensemble of CNN to further increase the classification performance.

Janssens et al. [29] have proposed using a frequency analysis as input to a CNN. Four different conditions (three different failure types and the intact state) are classified using the FFT data points of two acceleration sensors which are mounted on the bearing housing. For each condition, five bearings of the same type are used to generate the dataset for classification. Several numbers of feature maps and number of layers are tested. Lastly it is stated that a deep version of the proposed architecture is not beneficial in that use-case.

Jing et al. [17] have compared CNN-based approaches operating on raw time data, frequency analysis and time-frequency analysis as pre-processing. Different network architectures consisting of different numbers of convolutional and pooling layers are employed. The investigation was conducted based on two datasets. Both datasets consist of acceleration measurements of a gearbox housing with gears of different health states. Frequency-based data are found to work best with the proposed network. CNN architectures with fewer layers result in higher accuracies than using more layers. An increase of the input segment size results also in higher accuracies.

Gray-scale images have been employed by Wen et al. in [20] by representing each time step by a pixel with the relative signal amplitude as pixel strength. Those gray-scale images are classified using a CNN architecture that is based on the LeNet-5 architecture [30]. Their approach is tested on three different datasets, namely the CWRU bearing data set, a self-priming centrifugal pump dataset, and an axial piston hydraulic pump dataset. Other deep learning and machine learning methods such as Deep Belief Networks, Probabilistic Neural Networks, or Sparse Filter result in similar accuracies as the proposed approach.

Zhang et al. [31] have also performed pre-processing of time series vibration data by generating gray-scale images. A CNN consisting of two convolutional layers each followed by a sub-sampling layer is applied for classification. The approach is compared to using raw time signal data that are classified using a CNN and using FFT data points that are classified using a neural network.

Lu et al. [32] have proposed a nearly identical approach as in [31]. Gray-scale images are classified using a CNN with two convolutional and two sub-sampling layers. Some minor adaptions of the CNN training, such as greedy forward learning or a local connection between two layers, are proposed to increase robustness of the classification. In addition, the parameters of the convolutional and sub-sampling layers are different from [31]. To test the robustness, additional noise is added to the vibration sensor data. The proposed method achieves higher accuracy rates compared to a SVM or a shallow softmax regression classifier. However, a stacked de-noising Autoencoder results in a classification accuracy comparable to the proposed CNN approach.

Guo et al. [33] do not mention any pre-processing but transform time series data to a matrix which is in fact using a gray-scale image. The CNN for classification consists of three combinations of one convolutional and one sub-sampling layer followed by two fully connected layers.

Liao et al. [16] have compared WT and STFT as a pre-processing method for time series data. The classification is performed using a CNN that consists of two convolutional layers, each followed by a sub-sampling layer with a fully connected layer at the end of the network. Vibrational data of ten different health states of an automotive gearbox are recorded on a test bench. Using WT input data requires less training iterations compared to using STFT data.

Verstraete et al. [21] have analyzed STFT, WT, and Hilbert-Huang Transformation (HHT) as a pre-processing method for a classification using CNN. It is claimed that a STFT cannot represent transient signals adequately while WT is effective for transient signals. The HHT is said to be suited for in-stationary signals but has numerical problems resulting in negative values under specific circumstances of the time signal. The proposed network architecture consists of two consecutive convolutional layers followed by one pooling layer. This convolution/pooling-layer combination is repeated three times and then followed by two fully connected layers. The double convolution is said to reduce the number of parameters of the network and should improve the generated features due to the additional non-linearity. The approach is tested on two bearing datasets, one of which is the CWRU bearing data set. The average classification accuracy on both datasets using the WT is slightly higher than using the STFT and the HHT has the lowest accuracy.

Zhang et al. [34] have performed a STFT on the data of the CWRU and classified them using a similar network architecture as in [21]. It consists of two consecutive convolutional layers followed by a pooling layer. In general, the overall approach in [34] is quite similar to [21], which explains the similar result.

Wavelets have also been employed by Ding and He [23] to face varying operational conditions of bearing applications. They propose a signal-to-image-approach based on Wavelet Packet Transform. This representation is used in a customized CNN that combines features from a convolutional layer and a sub-sampling layer in a special multi-scale layer. The authors claim that it enables more invariant and robust features with precise details.

The periodicity of a time signal can be visualized using a Recurrence Plot (RP) that analyzes the signal's phase space trajectory. It reflects those points at the time at which trajectories in phase space return to a previous (or close-to-previous) state. The classification performance using a RP as pre-processing method is compared to seven other time series classification algorithms based on 20 real-world datasets in [35]. The RP approach results in the lowest error rate for 10 of 20 datasets.

Another image generating pre-processing method is the calculation of Gramian Angular Fields (GAFs). They were first introduced in [36,37] for encoding time series and are used by [38] to detect defects on railway wagon wheels using a CNN architecture.

Summarizing the state of the art does not give a clear suggestion for selecting a pre-processing method or a network configuration. Various methods (such as no pre-processing, gray-scale image, STFT, WT) result in a testing accuracy above 99 % when classifying the CWRU bearing data set of [25]. Therefore, in the next section, we screen several pre-processing methods.

3. Conceptual Analysis Approach

Applications of CNN show promising results in the area of machine health monitoring, as shown in Section 2. However, we can find neither a favorable pre-processing method, nor do we see clear suggestions in the literature for the choice of the network architecture or its hyper-parameters, such as spatial extent, the number of kernels, or the network depth. After a description of the dataset in Section 3.1, we therefore investigate various established pre-processing methods in Section 3.2. Promising pre-processing methods are selected as input data for a network architecture investigation in Section 3.3 concerning the size for the receptive field of the convolutional layers, the size of the pooling layer as well as the network depth. To further investigate the findings regarding the network architecture, Section 3.4 investigates the CNN's feature extraction by analyzing the trained kernel weights.

3.1. Description of the Dataset

The analysis in this paper is based on the actual driving data of an upper class sedan vehicle of a Bavarian manufacturer with a semi-active suspension system. Defective dampers are simulated by setting constant damper currents that lead to reduced damping forces. Figure 1 shows the characteristic curve of an intact and defective damper for each axle. Even though there is a great variety of different damper defects and related consequences, simulating defective dampers by changing damper currents is a reasonable approach according to [6,39].

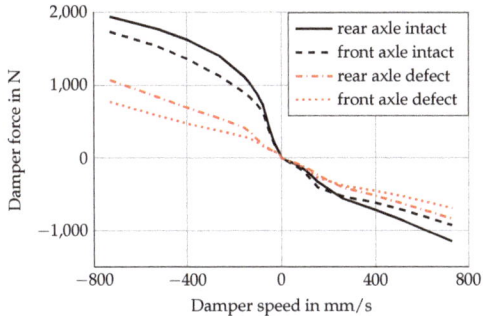

Figure 1. Characteristic curve of a single damper

Figure 2 visualizes the overall classification process. Although the measurement data of this paper were recorded using an upper class vehicle with semi-active suspension, it should be possible to apply the diagnosis approach to vehicles with a traditional passive chassis system. Therefore, only seven sensor signals from the vehicle's ESC system (four wheel speed signals, lateral and longitudinal accelerations as well as yaw rate) were utilized for our approach. Each sensor signal is logged with a sampling frequency of $f_s = 100$ Hz generating raw time signals. A sequence of 512 sequential data points is called an observation and each observation is categorized according to its damper health state. To comply with an average driving style, an observation is required to have an average longitudinal and lateral acceleration of less than 1 m/s^2 as well as to have an average speed above 30 km/h. The dataset consists of nearly 13,000 observations covering a distance of 1650 km which is around 18 h of driving on the German Autobahn, national and country roads as well as bad roads. The dataset is evenly distributed among the classes

- all dampers intact,
- all dampers defect,
- front left (FL) damper defect with other dampers intact and
- rear right (RR) damper defect with other dampers intact,

representing an intact suspension system, wear on all dampers due to aging, and two different single damper defects. The dataset is divided into 80 % training data and 20 % testing data. For a 5-fold cross-validation, the training data is further divided into 5 folds, whereas 4 folds are used for actual training and 1 fold is used as validation data.

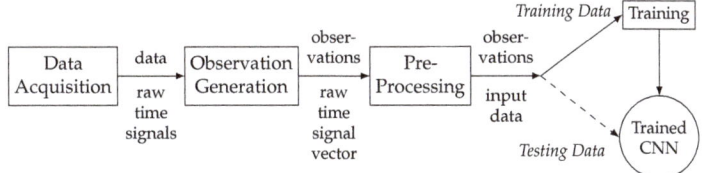

Figure 2. Overview of classification process

3.2. Analysis of Pre-Processing Methods

3.2.1. CNN Architecture for Pre-Processing Analysis

For the analysis of different pre-processing methods, a suitable CNN architecture needs to be defined. State of the art CNN architectures handle the first convolutional layer differently from the rest of the network. This stresses that the hyper-parameters of this layer should be chosen carefully. Furthermore, special building blocks for CNN have been proposed, e.g., Inception Modules [28,40] or Residual Connections [41–44]. At this point, we do not know how different kernel sizes affect the results. Therefore, we use a Inception Module-like block as the first layer. This enables the network to extract features on different scales and prevents any pre-processing from suffering under unsuitable network architectures, which would lead to a decrease in performance. The Inception Module is a powerful, yet complex building block we use in our CNN and does not limit the network to a specific kernel size. The building block is depicted in Figure 3.

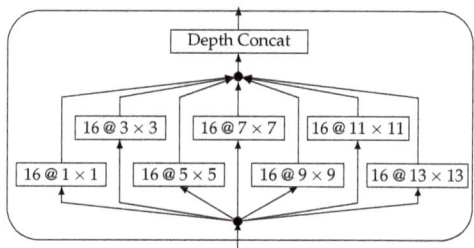

Figure 3. Inception-like module that is used as the first convolutional layer. The notation "16 @ 3 × 3" implies 16 filter kernels with a spatial extent of 3 × 3. All layers use *same* padding and a stride of 1.

We use seven branches with different kernel sizes and the same number of kernels for each branch. The stride is set to 1 and the padding of each convolution is chosen to be *same*, resulting in equally sized feature maps. This allows for the depth-wise concatenation of all extracted feature maps. The amount of filter kernels is chosen identically in order for there to be no one kernel size preferred over the other. Larger filters can learn lower frequencies from raw data than smaller filters, whereas small filters can be beneficial for processing peaks in the frequency spectrum data. The Inception-like module is integrated into the overall architecture given in Table 1.

Table 1. Architecture for the evaluation of different pre-processings

Layer	Details
Input	7 channels (4 wheel speeds, lat. & long. acceleration, yaw rate)
Inception-like module	See Figure 3
Max-Pooling	Kernel size 2 × 2, stride 2, *valid* padding
Fully connected	128 neurons
Dropout	Dropout rate 0.5
Output	4 neurons

Except for the Inception-like module, the architecture of the CNN for evaluating different pre-processing methods remains simple. Max-Pooling is commonly used to establish invariance to small local changes and reduce the amount of parameters, which is why we add a single sub-sampling layer. The feature extraction stage is followed by a fully connected layer and uses dropout [45] as a simple regularization technique. The overall network architecture is shallow. Therefore, we do not make use of Batch-Normalization [46,47] or Residual Connections [41–44], which can improve convergence and significantly improve training speed in deeper networks. To prevent our network from over-fitting, we employ L2-regularization. A hyper-parameter optimization to select learning rate and L2-regularization is conducted. The cost function of this optimization is the average of the validation accuracy and the difference of training and validation accuracy to prevent over-fitting. Further details of the implementation of the neural networks are described in Appendix A.

3.2.2. Description of Pre-Processing Methods

Looking at the state of the art of CNN applications for fault detection and isolation systems of mechanical components in Section 2, many different pre-processing techniques exist. While some publications aim at implementing end-to-end-systems, which operate on raw data, others choose simple or more complex transformations, e.g., scaling, denoising, or Fourier transformations. These transformations result in one- or two-dimensional data representations. The selected methods for investigation are chosen to represent a broad bandwidth regarding the pre-processing complexity and are explained in the following paragraphs. Figure 4 shows a data sample of the front left wheel speed n_{FL} that was processed with these methods.

A simple option for pre-processing is removing linear trends within the data samples. Because the driving data have been recorded at varying vehicle velocities, the magnitude of the wheel speed can be different. By subtracting a linear function, we aim at removing any bias or non-stationarity and focus primarily on transient dynamics within the signals. The linear detrend is applied before any of the other transformations we investigate.

To reduce the pre-processing effort as much as possible, simple scaling can be applied. This is recommended if the domain of multiple input channels scales differently as in this case, it speeds up the convergence of the commonly used back-propagation algorithm [48].

One-dimensional frequency-based data can be created by applying a FFT. A hanning window is applied and the one-sided spectrum is used as input data for the CNN. With a sampling frequency of 100 Hz, the maximum frequency of the FFT is 50 Hz and the input dimension is reduced from 512 data points to 256 data points per sensor signal.

Grayscale images from [20] are generated by reshaping the time signal vector to a matrix. The value of the very first data point of the time signal is indicated by the color in the top left corner and the very last data point of the time signal projected to the bottom right corner. Each row represents consecutive data points of the signal.

The STFT employs a Fourier Transformation to extract the frequency components of local sections (windows) of a signal as it changes over time. The width of the windowing function relates between frequency and time resolution. The choice of transformation parameters is decisive for the resulting size of the spectrogram. We choose a FFT segment length of 64 and an overlap of 8 samples. The resulting image is then transformed to a 32 × 32 pixel image. [17] suggests using a STFT rather than a combination of raw signal and FFT data. Therefore, it is employed by [16,21].

A GAF is the trigonometric sum between all points of a transformation of time series data to polar coordinates.

A phase-based two-dimensional representation, looking similar to GAFs, can be constructed by RPs [49]. The algorithm calculates a matrix norm of all data points (of a single sample) to each other and thus maps the time axis to a matrix. Hence, a signal of length n becomes an image of size $n \times n$. Due to memory requirements, both the GAFs and RPs are down-sampled to a size of 32 × 32.

The WT as used in [23] decomposes a signal by wavelet packets [50]. The energy of the signal is calculated based on the reconstructed coefficients of the wavelet packet nodes. The resulting energy vector is then modified to a two-dimensional image according to the phase space reconstruction technique from [51].

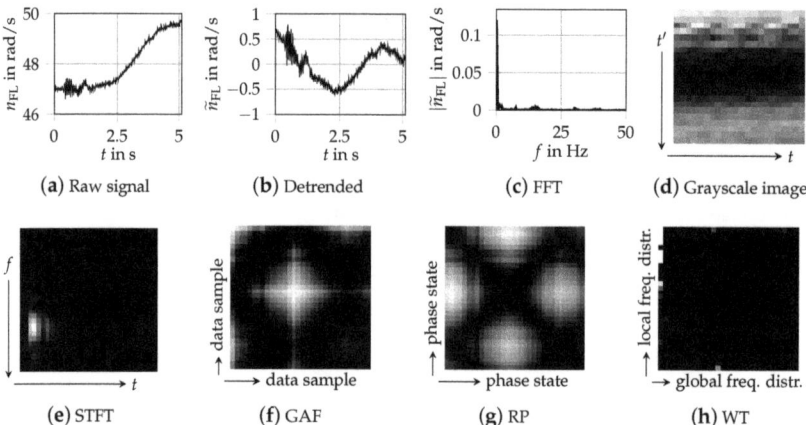

Figure 4. Different pre-processing representations of a data sample of the wheel speed front left n_{FL}.

The results for different pre-processing methods are given in Table 2. The used classification accuracy is the relation of the number of correct predictions divided by the number of observations. The statistical robustness is indicated by a mean accuracy and a standard deviation of the 5-fold cross-validation. Due to the balanced dataset consisting of four classes, randomly guessing results in 25% classification accuracy.

Table 2. Test data results of a 5-fold cross-validation for different pre-processing methods

Input Data	Pre-Processing Type	Classification Accuracy
1D	None (raw signal)	25.2 ± 0.4 %
	Detrending (remove linear trend)	79.4 ± 1.4 %
	Detrending and scaling to $[-1, 1]$	77.4 ± 1.4 %
	Detrending and scaling to Gaussian Distribution	81.2 ± 0.5 %
	Detrending and apply FFT	91.1 ± 0.2 %
2D	Detrending and apply Grayscale image	25.0 ± 0.0 %
	Detrending and apply STFT (32 × 32)	89.8 ± 0.2 %
	Detrending and apply GAF (32 × 32)	32.1 ± 1.8 %
	Detrending and apply RP (32 × 32)	47.1 ± 11.7 %
	Detrending and apply WT (32 × 32)	72.7 ± 1.0 %

Operating directly on the raw data does not enable a classification due to the different vehicle speeds of the dataset. Applying a linear detrend to the time signals already improves the accuracy to nearly 80%. Additional scaling changes this result only slightly. Applying a FFT results in the best classification accuracy. Generating gray-scale images behaves as poor as using raw data directly. The STFT results in the best classification accuracy of the two-dimensional input data versions, followed by the WT. The rather complex images generated by GAF and RP do not improve the classification. We further investigate "detrending", FFT, and STFT as a pre-processing method. Even though detrending results in a 12 percentage points (pps) lower accuracy compared to using FFT data, the performance of this simple pre-processing method using an optimized network architecture as well as the performed analyses of a CNN on time signals is of interest for the derivation of deeper knowledge of the CNN behavior.

3.3. Investigation of the Network Architecture

The investigation of the network architecture shall improve the classification performance and derive recommendations for the application of CNN to damper diagnosis and its hyper-parameter settings. The analysis is performed by varying the size of the receptive field e, the size of the max-pooling layer p and the depth of the network in terms of the number of consecutive convolutional layers d. A first screening shows that using 16 kernels in every convolutional layer is sufficient, as some kernels tend to have very small weights and therefore, do not make any contribution to the network's output. Figure 5 shows the classification accuracy of various network configurations for detrended (first column), FFT (second column), and STFT (third column) input data using $d = 1$ convolutional layer (first row) and $d = 3$ convolutional layers (second row).

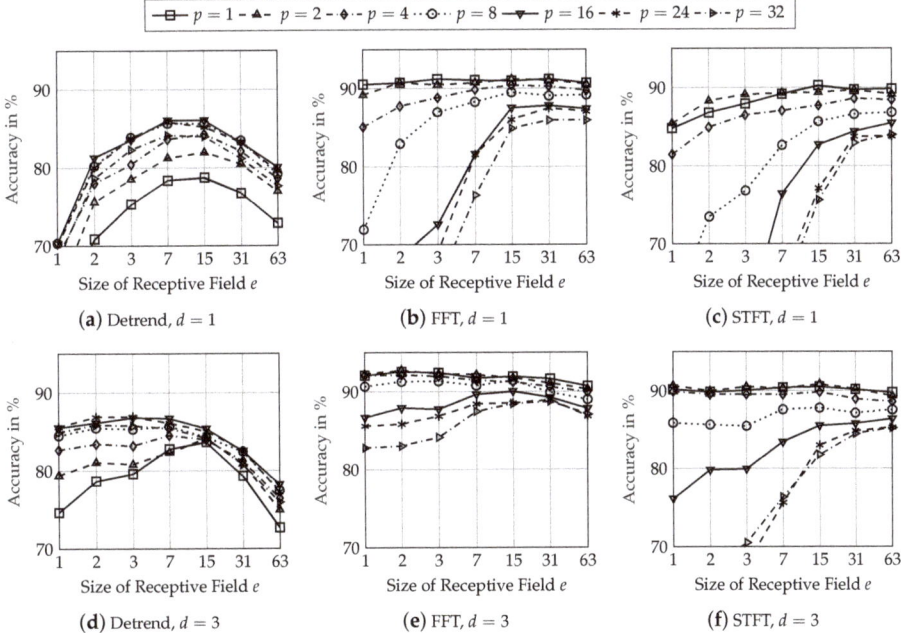

Figure 5. Mean test data accuracy of a 5-fold cross-validation for the investigation of the network architecture. p is the size of the pooling layer (with $p = 1$ effectively resulting in no pooling), e is the spatial extent of a kernel (for STFT the size of the receptive field is $e \times e$).

When using detrended input data, a greater receptive field ($e \geq 7$) is especially important. A receptive field of $e = 7$ and a data sampling frequency of 100 Hz corresponds to an oscillation frequency of approximately 14 Hz. This is approximately a chassis system's vertical eigenfrequency [52]. Model-based damper defect detection approaches also operate mainly based on this frequency [6]. Therefore, the size of the receptive field in combination with the data sampling frequency should be selected so that at least this eigenfrequency can be detected. A greater size of the pooling layer ($p \geq 8$) results in the best performance when using detrended input data. The pooling layer generates local invariance of the generated features of the preceding convolutional layer. Hereby, phase shifts between the input signal and kernel weights are compensated. This requires a size of the pooling layer of around $p = 8$ for the chassis system's vertical eigenfrequency.

When using FFT input data, the amplitude of the chassis system's vertical eigenfrequency is already included in the input data. For a small size of the pooling layer, the performance is nearly equal for different sizes of the receptive field. A pooling layer may be even disadvantageous for frequency analysis input data if the convolutional layer compares different amplitudes at different frequencies. In fact, the smallest size of pooling layer $p = 1$ (effectively resulting in no pooling) results in the highest accuracy.

Similarly to FFT input data, the amplitudes at different frequencies are available in the STFT input data with additional time-relation information. Therefore, the behavior of using STFT input data is similar to using FFT input data. Small sizes of the pooling layer result in the best accuracy. A greater size of the receptive field increases the classification performance, especially for large sizes of the pooling layer. Even a receptive field that is larger than the actual input data ($e = 63$) results in accuracies similar to those of smaller receptive fields. However, this increases the number of trainable parameters and is therefore of no further benefit.

Additional convolutional layers improve the performance of all network configurations for every pre-processing method. The accuracies of the best performing network architectures with one convolutional layer are increased by about 1 pp. Network architectures that result in a very low accuracy with one convolutional layer have a higher increase of their classification performance with $d = 3$ convolutional layers. Due to this behavior and as additional convolutional layers generate more abstract features, the robustness regarding unknown effects in testing data might be higher for deeper network architectures.

3.4. Investigation of Kernel Weights

This section investigates the assumptions from Section 3.3 regarding the analyses of the input data within the neural network. Therefore, the learned kernel weights are investigated. It is assumed that supplying time-related detrended input data results in a frequency analysis, while supplying frequency analysis input data such as a FFT leads to a comparison of amplitudes at different frequencies. Figure 6 visualizes two trained kernel weights for a selected network architecture for each pre-processing method. The network architectures with depth $d = 1$ resulting in the highest accuracy was selected, which is $d = 1$, $e = 15$ and $p = 16$ for detrended input data. For FFT input data, the selected network architecture is $d = 1$, $e = 63$ and $p = 1$ as a greater size of the receptive field allows for greater insight regarding the comparison of frequency amplitudes compared to a small size of the receptive field. For STFT input data, the network architecture $d = 1$, $e = 15$ and $p = 1$ was selected as this results in the highest accuracy and the size of the receptive field still enables the interpretation of the kernel weights.

Figure 6a,d shows the weights of two kernels out of 16 for a network with detrended input data. Both kernels learned weights to analyze oscillations of the input data at different frequencies. As the shape of the weights of one kernel is similar for all input signals, each input signal is analyzed for the same frequency by one kernel. With a sampling rate of the input data of 0.01 s, Figure 6a accounts for lower frequencies (below 7 Hz), while Figure 6d accounts for higher frequencies of above 15 Hz.

Figure 6b,e shows the weights of a network with FFT input data. The kernels calculate the weighted average of the signal amplitudes at the frequencies within the receptive field. As the shape of the kernel weights consists of distinct peaks, the kernels compare amplitudes at given frequencies. The sign of the weights only matters for the following activation function but has no further physical interpretation. Sliding these peaky kernels over the FFT input data, results in a comparison of the signal amplitudes at the frequency difference indicated by the distance of the peaks. The shape of the kernel weights is similar for all signals within one kernel. Therefore, amplitudes of the same frequency differences are compared for each signal by a specific kernel. With a frequency difference between two indices of the receptive field of $\Delta f = \frac{1}{512*0.01s} = 0.1953$ Hz, the kernel in Figure 6b analyzes for frequency differences of around $(37 - 13) * \Delta f = 4.7$ Hz and the kernel in Figure 6e analyzes for frequency differences of around $(46 - 9) * \Delta f = 7.2$ Hz.

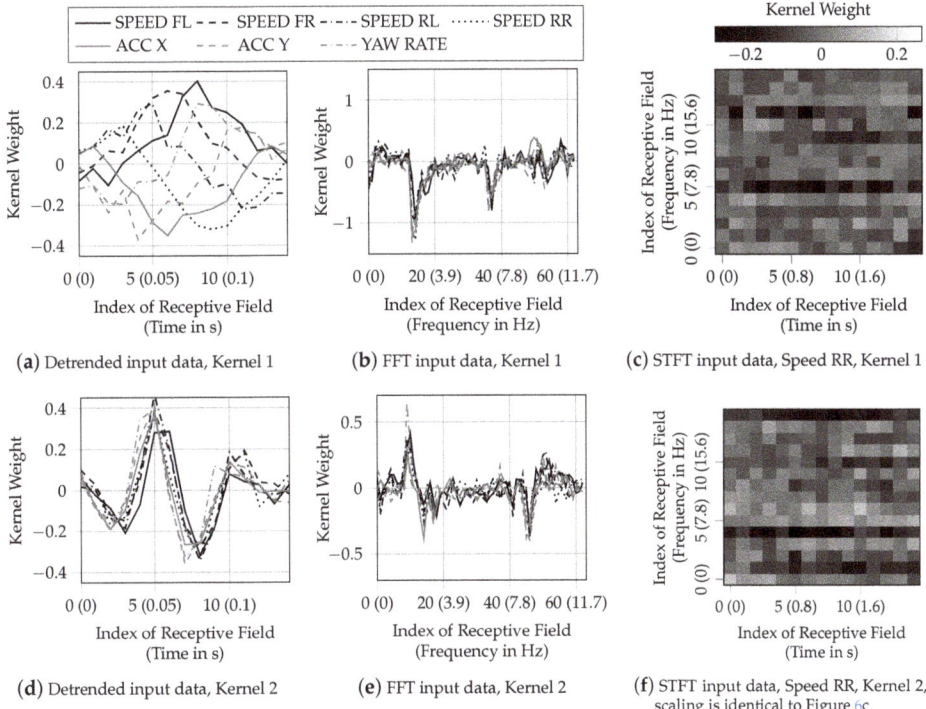

Figure 6. Kernel weights of the convolutional layer of different network architectures.

Two-dimensional input data results in two-dimensional kernels for each signal. Figure 6c,f visualizes the kernel weights for the rear right speed signal for a network with STFT input data. Aside from noise, noteworthy characteristics are dark horizontal lines. This demonstrates that this CNN performs analyses similar to the CNN using FFT input data by analyzing for amplitude differences at different frequencies independently from the time information on the x-axis. With the input data resolution of 32 pixels, the frequency resolution is $\Delta f = \frac{f_s}{2*32} = 1.56$ Hz. Therefore, the kernel in Figure 6c analyzes for frequency differences of $(11-5) * \Delta f = 9.4$ Hz and the kernel in Figure 6f analyzes for frequency differences of $(14-4) * \Delta f = 15.6$ Hz. While the frequency difference of 15.6 Hz has no obvious technical interpretation, 9.4 Hz are about the difference of the vehicle's body vertical eigenfrequency and the chassis's system vertical eigenfrequency. Because the time information is not used by the CNN, using a STFT is of no benefit compared to using FFT input data. It even reduces the resolution of the frequency analysis and therefore even might be disadvantageous for the classification task.

4. Results And Robustness

To evaluate the robustness of the trained classification systems, we created two additional different test datasets. We changed from summer to winter tires while leaving the rest of the vehicle setup unchanged. This dataset is called "tire variation". The vehicle setup of the second robustness test dataset, called "mass variation", consisted of an additional load of 200 kg in the trunk of the vehicle. Due to the package of the vehicle, this additional mass mainly affected the rear axle. Both datasets were gathered on the German Autobahn as well as national and country roads. The mass variation dataset consists of 1270 observations from 180 km. The driven roads are partly equal to the training dataset and partly different. The tire variation dataset consists of 2049 observations from 270 km, driven on completely different roads compared to the original test dataset from Section 3. Defective dampers were simulated identically as explained in Section 3.1. The distribution of the four defective classes was nearly balanced for both datasets. In the following, the networks were trained based on the training dataset from Section 3. Therefore, effects of changed tires or additional mass are completely unknown to the trained networks. Figure 7 shows the classification accuracies of the network architectures from Figure 5 applied to the mass variation dataset and Figure 8 shows the results for the tire variation dataset.

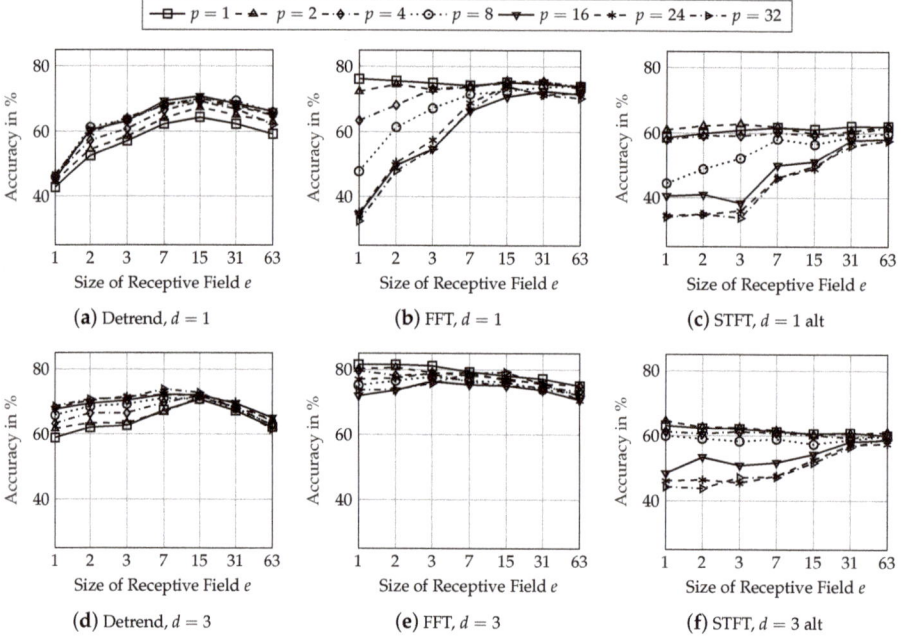

Figure 7. Mean accuracy of the 5-fold cross validation network architectures from Figure 5 applied to the mass variation dataset.

In general, the classification performance on both robustness datasets is lower than on the initial test dataset from Figure 5. Compared to the initial test dataset, the accuracy of the best network architecture using detrended input data is reduced by 13 pps on the mass dataset and by 23 pps on the tire dataset. When using FFT input data, the reduction of the accuracy is 11 pps on both the mass and tire dataset. Using STFT input data results in the highest reduction of classification accuracy. The accuracy on the mass dataset is 25 pps lower than on the initial test dataset and reduced by approximately 40 pps on the tire dataset.

Effects of the network architecture regarding the size of the receptive field e and the size of the pooling layer p are identical to the findings from Section 3.3. The accuracy is increased by about 1 pp for the best performing network architectures with an increasing depth of $d = 3$. When using FFT input data, the performance on the mass and tire datasets is increased by over 5 pps with an increase of the network depth when investigating the best network architectures from Section 3.3 ($e = 1$, $p = 1$ and $d = 1$). The classification performance using STFT input data is also increased with more convolutional layers but still remains at a low level of 50 and 60 %.

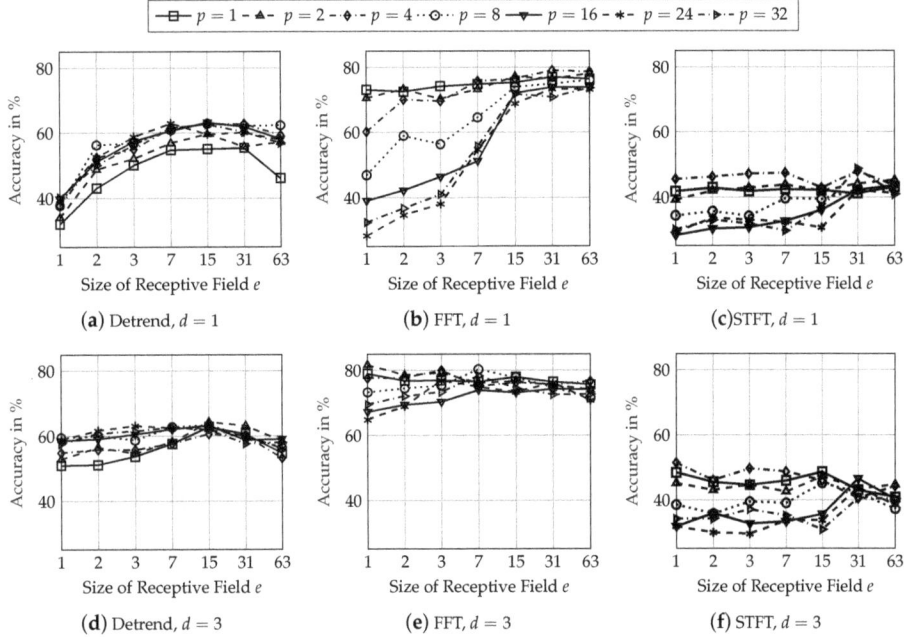

Figure 8. Mean accuracy of the 5-fold cross validation network architectures from Figure 5 applied to the tire variation dataset.

5. Discussion

The best performing network architectures using frequency analysis input data (FFT and STFT) require a small size of the pooling layer as well as a small size of the receptive field. The extreme case of $e = 1$ and $p = 1$ even results in a simple scaling of the frequency analysis of each sensor signal and averaging across the data points of all sensor signals. The actual classification is performed in the fully connected layer using those averaged frequency analysis data points. This raises the question of whether a CNN is necessary or if a Multi-Layer Perceptron (MLP) neural network is sufficient for solving the classification task. Therefore, a MLP neural network consisting of only a fully connected layer with 128 neurons was trained using the FFT input data. The resulting classification accuracy on the test dataset was 87.27 %, 72.97 % on the mass dataset and 68.37 % on the tire dataset. This demonstrates that the convolutional operation adds robustness to the performance of the neural network.

While promising results for diagnosing defective dampers using Convolutional Neural Networks are presented in this paper, limitations for a real-world implementation still exist, which will be discussed in the following paragraphs.

There are only data of one specific vehicle used in this paper's investigations. All networks were trained and tested using data of this vehicle. Recording training data for every unique vehicle configuration with different damper defects during the development phase seems challenging for an actual implementation. Therefore, portability of a trained classification system with high generalization capabilities for the diagnosis of different vehicles is desirable.

Robustness is a critical aspect for a real-world application. Even though the network architectures showed a robust behavior for tire and mass variations, further robustness analysis is necessary because there is a great variety of different circumstances during the usage of a vehicle.

Real damper defects might not occur in a switching manner, but the loss of damping forces might increase gradually over a long period of time. A classification process for the damper's health state might therefore not detect a minor defect. This can be encountered in two ways: Adding additional classes or predicting a continuous score for the health state of each damper. However, both approaches raise the need of additional training data.

6. Conclusions

This paper analyzed the suitability of using Convolutional Neural Networks for the diagnosis of automotive damper defects using driving data of the longitudinal and lateral acceleration as well as yaw rate and wheel speeds. The classification performance using different pre-processing methods was analyzed. Using detrended time-signals as well as frequency analyses such as FFT or STFT showed the best results.

The analysis of the network architecture showed that the size of the receptive field and the size of the pooling layer needs to be chosen according to relevant oscillation frequencies of the input signal when using time-related detrended input data. The analysis of the trained kernel weights demonstrated that a frequency analysis is performed by the CNN for detrended time-signal input data.

Using FFT input data results in the overall best classification performance. A small size of the pooling layer performs best and the size of the receptive field can be chosen arbitrarily. The trained kernels perform a comparison of the amplitudes for several frequency differences.

Using STFT input data results in a similar classification performance as using FFT input data. The investigation of the trained kernels showed that the time information is not used by the CNN. Therefore, the STFT pre-processing does not result in any benefit. The reduced frequency resolution compared to a FFT pre-processing even decreased the robustness regarding unknown characteristics in the testing data such as additional mass or changed tires.

Table 3 shows the performance of the best network configurations of the three investigated pre-processing methods. The number of Floating Point Operations (FLOPs) for the execution of the model as well as the number of tuneable parameters are an indicator of the possibility of an implementation on the Electronic Control Unit (ECU) of a vehicle. However, since a specific value for the computing power of an automotive ECU cannot be found, the authors are not able to judge about the real-time implementation. The selected network configurations were chosen with regard to the performance of the three different datasets. The best classification accuracy results from using FFT input data with less network parameters than when using STFT input data.

The software and data of this paper are available online [53] (see the Supplementary Materials).

Table 3. Comparison of best performing network configurations

Pre-Processing	Network Architecture	Tuneable Parameters	Number of Model FLOPs	Mean Accuracy in %		
				Test	Mass	Tire
Detrended Time-Signals	$e=15, p=16, d=3$	69.444	485k	85.40	72.08	63.21
Detrending with a FFT	$e=1, p=2, d=3$	264.484	3.67M	92.22	80.14	81.65
Detrending with a STFT	$e=1, p=1, d=3$	2.102.564	14.7M	90.17	63.17	48.40

Supplementary Materials: The software and data is available online at [53]. https://github.com/TUMFTM/Damper-Defect-Detection-Using-CNN/.

Author Contributions: T.Z. (corresponding author) initiated this paper. His contribution to the overall methodology of the proposed approach was essential and he performed the analyses presented in this paper. T.Z. supervised the master thesis of T.H.-P. who made essential contributions to the implementation of the proposed approaches. Both implemented this paper. M.L. made an essential contribution to the conception of the research project. He revised the paper critically for important intellectual content. M.L. gave final approval of the version to be published and agrees to all aspects of the work. As a guarantor, he accepts responsibility for the overall integrity of the paper. Conceptualization, T.Z.; methodology, T.Z. and T.H-P.; software, T.Z. and T.H-P.; validation, T.Z. and T.H-P.; formal analysis,T.Z. and T.H-P.; investigation, T.Z. and T.H-P.; resources, T.Z.; data curation, T.Z.; writing–original draft preparation, T.Z. and T.H-P.; writing–review and editing, T.Z.; visualization, T.Z. and T.H-P.; supervision, T.Z. and M.L.; project administration, T.Z.; funding acquisition, M.L. All authors have read and agreed to the published version of the manuscript.

Funding: The work described in this paper was supported by the basic research fund of the Institute of Automotive Technology from the Technical University of Munich. This work was supported by the German Research Foundation (DFG) and the Technical University of Munich (TUM) in the framework of the Open Access Publishing Program.

Acknowledgments: We would like to thank the vehicle manufacturer for supplying us with a research vehicle.

Conflicts of Interest: The authors declare no conflict of interest.

Appendix A. Implementation Details

The CNN were implemented in TensorFlow 1.9 using Python 3.5. During training, we use Adam optimizer [54] with a mini-batch size of 128. The optimizer minimizes the sum of a Sparse Softmax Cross Entropy plus the sum of weight decay (L2 regularization). Training is stopped after 650 epochs or if the accuracy with validation data has not significantly improved for 50 epochs (Early Stopping). Kernels and weights are initially set following He Initialization scheme [55] and the bias terms are initially set to a small constant (0.01). We use Rectified Linear Unit (ReLU) activation functions for every layer except for the output units, which are linear neurons.

References

1. Kothamasu, R.; Huang, S.H.; VerDuin, W.H. System health monitoring and prognostics—A review of current paradigms and practices. *Int. J. Adv. Manuf. Technol.* **2006**, *28*, 1012–1024, doi:10.1007/s00170-004-2131-6. [CrossRef]
2. Ferreira, C.; Ventura, P.; Morais, R.; Valente, A.L.; Neves, C.; Reis, M.C. Sensing methodologies to determine automotive damper condition under vehicle normal operation. *Sens. Actuators Phys.* **2009**, *156*, 237–244, doi:10.1016/j.sna.2009.03.035. [CrossRef]
3. Hernandez-Alcantara, D.; Amezquita-Brooks, L.; Vivas-Lopez, C.; Morales-Menendez, R.; Ramirez-Mendoza, R. Fault detection for automotive semi-active dampers. In Proceedings of the 2013 Conference on Control and Fault-Tolerant Systems (SysTol), Nice, France, 9–11 October 2013; pp. 625–630, doi:10.1109/SysTol.2013.6693916. [CrossRef]
4. Hernandez-Alcantara, D.; Morales-Menendez, R.; Amezquita-Brooks, L. Fault Detection for Automotive Shock Absorber. *J. Phys. Conf. Ser.* **2015**, *659*, 012037, doi:10.1088/1742-6596/659/1/012037. [CrossRef]
5. Alcantara, D.H.; Morales-Menendez, R.; Amezquita-Brooks, L. Fault diagnosis for an automotive suspension using particle filters. In Proceedings of the 2016 European Control Conference (ECC), Aalborg, Denmark, 29 June–1 July 2016; pp. 1898–1903, doi:10.1109/ECC.2016.7810568. [CrossRef]
6. Jautze, M. *Ein Signalmodellbasiertes Verfahren zum Erkennen von Dämpferschäden bei Kraftfahrzeugen: Zugl.: Erlangen-Nürnberg, Univ., Diss., 2002*; Fortschritt-Berichte VDI Reihe 12, Verkehrstechnik/Fahrzeugtechnik; VDI-Verl.: Düsseldorf, Germany, 2002; Volume 498.
7. Yin, S.; Li, X.; Gao, H.; Kaynak, O. Data-Based Techniques Focused on Modern Industry: An Overview. *IEEE Trans. Ind. Electron.* **2015**, *62*, 657–667, doi:10.1109/TIE.2014.2308133. [CrossRef]
8. Zehelein, T.; Merk, A.; Lienkamp, M. Damper diagnosis by artificial intelligence. In *9th International Munich Chassis Symposium 2018*; Pfeffer, P., Ed.; Springer Fachmedien Wiesbaden: Wiesbaden, Germany, 2019; pp. 461–482, doi:10.1007/978-3-658-22050-1_31. [CrossRef]
9. Zehelein, T.; Werk, P.; Lienkamp, M. An Evaluation of Autoencoder and Sparse Filter as Automated Feature Extraction Process for Automotive Damper Defect Diagnosis. In Proceedings of the 2019 Fourteenth International Conference on Ecological Vehicles and Renewable Energies (EVER), Monte-Carlo, Monaco, 8–10 May 2019; pp. 1–8, doi:10.1109/EVER.2019.8813630. [CrossRef]
10. Zhao, G.; Zhang, G.; Ge, Q.; Liu, X. Research advances in fault diagnosis and prognostic based on deep learning. In Proceedings of the 2016 Prognostics and System Health Management Conference (PHM-Chengdu), Chengdu, China, 19–21 October 2016; pp. 1–6, doi:10.1109/PHM.2016.7819786. [CrossRef]
11. Zhao, R.; Yan, R.; Chen, Z.; Mao, K.; Wang, P.; Gao, R.X. Deep learning and its applications to machine health monitoring. *Mech. Syst. Signal Process.* **2019**, *115*, 213–237, doi:10.1016/j.ymssp.2018.05.050. [CrossRef]

12. Sateesh Babu, G.; Zhao, P.; Li, X.L. Deep Convolutional Neural Network Based Regression Approach for Estimation of Remaining Useful Life. In *Database Systems for Advanced Applications*; Navathe, S.B., Wu, W., Shekhar, S., Du, X., Wang, X.S., Xiong, H., Eds.; Lecture Notes in Computer Science; Springer International Publishing: Cham, Switzerland, 2016; Volume 9642, pp. 214–228, doi:10.1007/978-3-319-32025-0_14. [CrossRef]
13. Li, X.; Ding, Q.; Sun, J.Q. Remaining useful life estimation in prognostics using deep convolution neural networks. *Reliab. Eng. Syst. Saf.* **2018**, *172*, 1–11, doi:10.1016/j.ress.2017.11.021. [CrossRef]
14. Zhang, W.; Li, R.; Deng, H.; Wang, L.; Lin, W.; Ji, S.; Shen, D. Deep convolutional neural networks for multi-modality isointense infant brain image segmentation. *NeuroImage* **2015**, *108*, 214–224, doi:10.1016/j.neuroimage.2014.12.061. [CrossRef]
15. Zahia, S.; Sierra-Sosa, D.; Garcia-Zapirain, B.; Elmaghraby, A. Tissue classification and segmentation of pressure injuries using convolutional neural networks. *Comput. Methods Progr. Biomed.* **2018**, *159*, 51–58, doi:10.1016/j.cmpb.2018.02.018. [CrossRef] [PubMed]
16. Liao, Y.; Zeng, X.; Li, W. Wavelet transform based convolutional neural network for gearbox fault classification. In Proceedings of the 2017 Prognostics and System Health Management Conference (PHM-Harbin), Harbin, China, 9–12 July 2017; pp. 1–6, doi:10.1109/PHM.2017.8079274. [CrossRef]
17. Jing, L.; Zhao, M.; Li, P.; Xu, X. A convolutional neural network based feature learning and fault diagnosis method for the condition monitoring of gearbox. *Measurement* **2017**, *111*, 1–10, doi:10.1016/j.measurement.2017.07.017. [CrossRef]
18. Zhang, W.; Li, C.; Peng, G.; Chen, Y.; Zhang, Z. A deep convolutional neural network with new training methods for bearing fault diagnosis under noisy environment and different working load. *Mech. Syst. Signal Process.* **2018**, *100*, 439–453, doi:10.1016/j.ymssp.2017.06.022. [CrossRef]
19. Appana, D.K.; Ahmad, W.; Kim, J.M. Speed Invariant Bearing Fault Characterization Using Convolutional Neural Networks. In *Multi-disciplinary Trends in Artificial Intelligence*; Phon-Amnuaisuk, S., Ang, S.P., Lee, S.Y., Eds.; Lecture Notes in Computer Sciencel; Springer International Publishing: Cham, Switzerland, 2017; Volume 10607, pp. 189–198, doi:10.1007/978-3-319-69456-6_16. [CrossRef]
20. Wen, L.; Li, X.; Gao, L.; Zhang, Y. A New Convolutional Neural Network-Based Data-Driven Fault Diagnosis Method. *IEEE Trans. Ind. Electron.* **2018**, *65*, 5990–5998, doi:10.1109/TIE.2017.2774777. [CrossRef]
21. Verstraete, D.; Ferrada, A.; Droguett, E.L.; Meruane, V.; Modarres, M. Deep Learning Enabled Fault Diagnosis Using Time-Frequency Image Analysis of Rolling Element Bearings. *Shock Vib.* **2017**, *2017*, 1–17, doi:10.1155/2017/5067651. [CrossRef]
22. Jia, F.; Lei, Y.; Lu, N.; Xing, S. Deep normalized convolutional neural network for imbalanced fault classification of machinery and its understanding via visualization. *Mech. Syst. Signal Process.* **2018**, *110*, 349–367, doi:10.1016/j.ymssp.2018.03.025. [CrossRef]
23. Ding, X.; He, Q. Energy-Fluctuated Multiscale Feature Learning With Deep ConvNet for Intelligent Spindle Bearing Fault Diagnosis. *IEEE Trans. Instrum. Meas.* **2017**, *66*, 1926–1935, doi:10.1109/TIM.2017.2674738. [CrossRef]
24. Xia, M.; Li, T.; Xu, L.; Liu, L.; de Silva, C.W. Fault Diagnosis for Rotating Machinery Using Multiple Sensors and Convolutional Neural Networks. *IEEE/ASME Trans. Mechatron.* **2018**, *23*, 101–110, doi:10.1109/TMECH.2017.2728371. [CrossRef]
25. Smith, W.A.; Randall, R.B. Rolling element bearing diagnostics using the Case Western Reserve University data: A benchmark study. *Mech. Syst. Signal Process.* **2015**, *64-65*, 100–131, doi:10.1016/j.ymssp.2015.04.021. [CrossRef]
26. Eren, L.; Ince, T.; Kiranyaz, S. A Generic Intelligent Bearing Fault Diagnosis System Using Compact Adaptive 1D CNN Classifier. *J. Signal Process. Syst.* **2019**, *91*, 179–189, doi:10.1007/s11265-018-1378-3. [CrossRef]
27. Zilong, Z.; Wei, Q. Intelligent fault diagnosis of rolling bearing using one-dimensional multi-scale deep convolutional neural network based health state classification. In Proceedings of the ICNSC 2018, Zhuhai, China, 27–29 March 2018; pp. 1–6, doi:10.1109/ICNSC.2018.8361296. [CrossRef]
28. Szegedy, C.; Liu, W.; Jia, Y.; Sermanet, P.; Reed, S.; Anguelov, D.; Erhan, D.; Vanhoucke, V.; Rabinovich, A. Going Deeper with Convolutions. *arXiv* **2014**, arXiv:1409.4842.
29. Janssens, O.; Slavkovikj, V.; Vervisch, B.; Stockman, K.; Loccufier, M.; Verstockt, S.; van de Walle, R.; van Hoecke, S. Convolutional Neural Network Based Fault Detection for Rotating Machinery. *J. Sound Vib.* **2016**, *377*, 331–345, doi:10.1016/j.jsv.2016.05.027. [CrossRef]

30. Lecun, Y.; Bottou, L.; Bengio, Y.; Haffner, P. Gradient-based learning applied to document recognition. *Proc. IEEE* **1998**, *86*, 2278–2324, doi:10.1109/5.726791. [CrossRef]
31. Zhang, W.; Peng, G.; Li, C. Bearings Fault Diagnosis Based on Convolutional Neural Networks with 2-D Representation of Vibration Signals as Input. *MATEC Web Conf.* **2017**, *95*, 13001, doi:10.1051/matecconf/20179513001. [CrossRef]
32. Lu, C.; Wang, Z.; Zhou, B. Intelligent fault diagnosis of rolling bearing using hierarchical convolutional network based health state classification. *Adv. Eng. Inform.* **2017**, *32*, 139–151, doi:10.1016/j.aei.2017.02.005. [CrossRef]
33. Guo, X.; Chen, L.; Shen, C. Hierarchical adaptive deep convolution neural network and its application to bearing fault diagnosis. *Measurement* **2016**, *93*, 490–502, doi:10.1016/j.measurement.2016.07.054. [CrossRef]
34. Zhang, W.; Zhang, F.; Chen, W.; Jiang, Y.; Song, D. Fault State Recognition of Rolling Bearing Based Fully Convolutional Network. *Comput. Sci. Eng.* **2018**, 1. doi:10.1109/MCSE.2018.110113254. [CrossRef]
35. Hatami, N.; Gavet, Y.; Debayle, J. Classification of Time-Series Images Using Deep Convolutional Neural Networks. *arXiv* **2017**, arXiv:1710.00886.
36. Wang, Z.; Oates, T. Encoding Time Series as Images for Visual Inspection and Classification Using Tiled Convolutional Neural Networks. In *Trajectory-Based Behavior Analytics*; Technical report/Association for the Advancement of Artificial Intelligence WS; AAAI Press: Palo Alto, CA, USA, 2015; pp. 40–46.
37. Wang, Z.; Oates, T. Imaging Time-Series to Improve Classification and Imputation. In *Proceedings of the Twenty-Fourth International Joint Conference on Artificial Intelligence*; Yang, Q., Wooldridge, M.J., Eds.; AAAI Press International Joint Conferences on Artificial Intelligence: Palo Alto, CA, USA, 2015.
38. Krummenacher, G.; Ong, C.S.; Koller, S.; Kobayashi, S.; Buhmann, J.M. Wheel Defect Detection With Machine Learning. *IEEE Trans. Intell. Transp. Syst.* **2018**, *19*, 1176–1187, doi:10.1109/TITS.2017.2720721. [CrossRef]
39. Hernández-Alcántara, D.; Tudón-Martínez, J.C.; Amézquita-Brooks, L.; Vivas-López, C.A.; Morales-Menéndez, R. Modeling, diagnosis and estimation of actuator faults in vehicle suspensions. *Control. Eng. Pract.* **2016**, *49*, 173–186, doi:10.1016/j.conengprac.2015.12.002. [CrossRef]
40. Szegedy, C.; Vanhoucke, V.; Ioffe, S.; Shlens, J.; Wojna, Z. Rethinking the Inception Architecture for Computer Vision. *arXiv* **2015**, arXiv:1512.00567
41. He, K.; Zhang, X.; Ren, S.; Sun, J. Deep Residual Learning for Image Recognition. *arXiv* **2015**, arXiv:1512.03385.
42. Szegedy, C.; Ioffe, S.; Vanhoucke, V.; Alemi, A. Inception-v4, Inception-ResNet and the Impact of Residual Connections on Learning. *arXiv* **2016**, arXiv:1602.07261
43. He, K.; Zhang, X.; Ren, S.; Sun, J. Identity Mappings in Deep Residual Networks. *arXiv* **2016**, arXiv:1603.05027.
44. Xie, S.; Girshick, R.; Dollár, P.; Tu, Z.; He, K. Aggregated Residual Transformations for Deep Neural Networks. *arXiv* **2016**, arXiv:1611.05431.
45. Srivastava, N.; Hinton, G.E.; Krizhevsky, A.; Sutskever, I.; Salakhutdinov, R. Dropout: A simple way to prevent neural networks from overfitting. *J. Mach. Learn. Res.* **2014**, *15*, 1929–1958.
46. Ioffe, S.; Szegedy, C. Batch Normalization: Accelerating Deep Network Training by Reducing Internal Covariate Shift. *arXiv* **2015**, arXiv:1502.03167.
47. Ioffe, S. Batch Renormalization: Towards Reducing Minibatch Dependence in Batch-Normalized Models. *arXiv* **2017**, arXiv:1702.03275.
48. LeCun, Y.; Bottou, L.; Orr, G.B.; Müller, K.R. Efficient BackProp. In *Neural Networks: Tricks of the Trade*; Goos, G., Hartmanis, J., van Leeuwen, J., Orr, G.B., Müller, K.R., Eds.; Lecture Notes in Computer Science; Springer Berlin Heidelberg: Berlin/Heidelberg, Germany, 1998; Volume 1524, pp. 9–50, doi:10.1007/3-540-49430-8_2. [CrossRef]
49. Marwan, N.; Carmen Romano, M.; Thiel, M.; Kurths, J. Recurrence plots for the analysis of complex systems. *Phys. Rep.* **2007**, *438*, 237–329, doi:10.1016/j.physrep.2006.11.001. [CrossRef]
50. Graps, A. An introduction to wavelets. *IEEE Comput. Sci. Eng.* **1995**, *2*, 50–61, doi:10.1109/99.388960. [CrossRef]
51. Kennel.; Brown.; Abarbanel. Determining embedding dimension for phase-space reconstruction using a geometrical construction. *Phys. Rev. At. Mol. Opt. Phys.* **1992**, *45*, 3403–3411, doi:10.1103/physreva.45.3403. [CrossRef]

52. Tuononen, A.; Hartikainen, L.; Petry, F.; Westermann, S. Parameterization of in-plane rigid ring tire model from instrumented vehicle measurements. In Proceedings of the 11th International Symposium on Advanced Vehicle Control (AVEC'12), Seoul, Korea, 9–12 September 2012.
53. Zehelein, T.; Hemmert-Pottmann, T. Damper Defect Detection Using CNN, 2019. Available online: https://github.com/TUMFTM/Damper-Defect-Detection-Using-CNN/ (accessed on 11 December 2019).
54. Kingma, D.P.; Ba, J. Adam: A Method for Stochastic Optimization. *arXiv* **2014**, arXiv:1412.6980.
55. He, K.; Zhang, X.; Ren, S.; Sun, J. Delving Deep into Rectifiers: Surpassing Human-Level Performance on ImageNet Classification. *arXiv* **2015**, arXiv:1502.01852.

© 2020 by the authors. Licensee MDPI, Basel, Switzerland. This article is an open access article distributed under the terms and conditions of the Creative Commons Attribution (CC BY) license (http://creativecommons.org/licenses/by/4.0/).

Article

V2X Communications Applied to Safety of Pedestrians and Vehicles

Fabio Arena [1], Giovanni Pau [1,*] and and Alessandro Severino [2]

1. Faculty of Engineering and Architecture, Kore University of Enna, 94100 Enna, Italy; fabio.arena@unikore.it
2. Department of Civil Engineering and Architecture, University of Catania, 95123 Catania, Italy; alessandro.severino@unict.it
* Correspondence: giovanni.pau@unikore.it

Received: 5 December 2019; Accepted: 23 December 2019; Published: 27 December 2019

Abstract: Connected cars and vehicle-to-everything (V2X) communication scenarios are attracting more researchers. There will be numerous possibilities offered by V2X in the future. For instance, in the case of vehicles that move in a column, they could react to the braking of those in front of it through the rapid information exchanges, and most chain collisions could be avoided. V2X will be desiderated for routes optimizations, travel time reductions, and accident rate decrease in cases such as communication with infrastructures, traffic information exchanges, functioning of traffic lights, possible situations of danger, and the presence of construction sites or traffic jams. Furthermore, there could be massive conversations between smartphones and vehicles performing real-time dialogues. It is relatively reasonable to expect a connection system in which a pedestrian can report its position to all surrounding vehicles. Regarding this, it is compelling to perceive the positive effects of the driver being aware of the presence of pedestrians when vehicles are moving on the roads. This paper introduces the concepts for the development of a solution based on V2X communications aimed at vehicle and pedestrian safety. A potential system architecture for the development of a real system, concerning the safety of vehicles and pedestrians, is suggested to draft some guidelines that could be followed in new applications.

Keywords: vehicle-to-everything communication; pedestrian; vehicles; safety

1. Introduction

In the last few years, a swelling phenomenon of crashes between vehicles and pedestrians [1], which has induced many injuries and even more deaths, has been witnessed [2,3]. Pedestrians, cyclists, and other non-vehicle occupants made up a considerable percentage of people killed in motor vehicle traffic crashes. Notwithstanding a constant lessening in the number of roadway fatalities, the number of pedestrian and cyclist fatalities has been nearly constant. As a consequence, these vulnerable groups have unique safety requirements and challenges that must be addressed. Thus, the protection of pedestrians, cyclists, and other non-vehicle occupants is part of the research areas regarding connected and automated vehicles [4]. In most cases, pedestrians collide with vehicles due to the distraction caused by their interactions with smartphones [5] or other Internet of things (IoT) devices [6–8]. This circumstance determines a decrease in the attention threshold and causes hazardous neglects during the crossing of roads [9–11]. Although there are some traditional mechanisms to protect pedestrians from vehicles [12], most of these are fundamentally linked to a pedestrian's acoustic warning [13]. However, this solution is often not adequate to divert the pedestrians' attentions from their smartphones. Pedestrian detection systems can be implemented in vehicles, in the infrastructure, or with pedestrians themselves to give warnings to drivers, pedestrians, or both. In-vehicle warning methods are becoming more and more conventional, for instance, blind-spot warning and forward collision warning. The contemporary field of vehicle-to-vehicle (V2V) communications is affording the advancement of high-level warning systems, such as intersection movement assist or left turn assist. In-vehicle warnings to the neighborhood of a pedestrian on the road might be logical. On the contrary, reasonably, the most straightforward and obvious warning system for pedestrians is a handheld device.

Fortunately, smartphones are becoming increasingly present and intelligent in people's daily lives [14,15]. It is also true that particular applications have been developed to provide proper alarms for pedestrians by using vehicle-to-pedestrian (V2P) communications [16,17]. In this paper, the concepts for the development of an approach based on V2X communications aiming at vehicle and pedestrian safety is proposed. The primary purpose of this work is not to present a solution that has already been implemented and tested. For this reason, performance evaluation is not carried out. Instead, this article aims to introduce the fundamental notions that a potential application for driver and pedestrian safety should support. To this end, we propose a possible hardware and software architecture that could be used in vehicles and mobile devices.

In this research field, the approach suggested in ref. [18] consists of an on-board unit (OBU) system designed for vehicles, which uses the IEEE 802.11p [19] standard in agreement with Wi-Fi protocols. The IEEE 802.11p protocol is an evolution of the IEEE 802.11 family, which introduces appropriate improvements aimed at supporting intelligent transport system (ITS) applications [20–24]. A more detailed validation of the OBU allows the realizing of a reliable V2V and V2P communication [25,26], both concerning the package delivery speed and the average delay [27]. Besides, in the literature, several security applications have been proposed to protect the safety of vehicles and pedestrians based on data transferred from the OBU. For instance, the solution introduced in refs. [28,29] is designed for showing driving information and providing a collision warning alarm on a tablet inside the vehicle. On the contrary, the application presented in refs. [13,30] has been developed for the smartphone that is supposed to be owned by the pedestrian. In this way, vital warning information can be provided to vulnerable pedestrians distracted by the smartphone.

Different solutions have been developed, aimed at interacting with the particular action that the smartphone is playing in a precise moment, i.e., listening to music, making calls, sending text messages. Meanwhile, the privacy protection is also crucial for all participants during the data exchange processes [31,32]. Based on the precise moment, it will be possible to appropriately manipulate four specific parameters that will allow interacting in the best way with the pedestrian. In particular, these parameters are aimed at establishing:

- The screen state;
- the item state;
- the voice screen state;
- the silent state of the smartphone.

The smart tuning and management of these parameters and related applications can report accidents near road intersections, and pedestrians can receive more appropriate warnings, based on the use of their smartphone. The main contributions of the concepts introduced in this paper are:

- Design of a V2X communication system to provide interfaces for vehicle and pedestrian in safety-oriented applications;
- proposal of two applications focused on vehicle and pedestrian safety, respectively, on tablets and smartphones.

This paper is organized as follows. Section 2 introduces a possible system architecture that can be used in the development of a safety application for vehicles and pedestrians, while specific concepts are analyzed in Section 3. Finally, Section 4 concludes the paper.

2. System Architecture

The possible hardware architecture of the system is depicted in Figure 1. It is composed of three components: The OBU installed on the vehicle, the tablet inside the vehicle, and the pedestrian's smartphone. Modern vehicles incorporate a vast range of electronic components. Consequently, it would not be challenging to include ad hoc-created OBUs or tablets. In fact, concerning the latter, almost all cars now include touchscreen that allow various operations to be carried out, such as, for instance, interaction with the audio or infotainment system, air conditioning management and vehicle parameters, and mirroring of different operating systems for mobile devices (iOS, Apple CarPlay; Android, Android Auto). Consequently, the implementation of ad hoc applications should not involve particularly complex efforts. Regarding the OBUs, they can be composed of multiple wireless interfaces, which permit the vehicles to communicate both with other vehicles and with roadside units. Considering the proposed architecture, the OBU includes a single-board computer, DSRC wireless interface (i.e., IEEE 802.11p), Wi-Fi interface (i.e., IEEE 802.11b/g/n/ac), cellular interface, GPS (global positioning system) receiver, and antennas for each technology, for instance, the round antenna for Wi-Fi and the rectangular antenna for IEEE 802.11p. The OBU is designed for interconnecting the different subsystems. Vehicle information can be collected by connecting the on-board diagnostic (OBD) interface and GPS. Information on surrounding vehicles can be collected through the IEEE 802.11p protocol. Smartphones and tablets can communicate with the OBU via the Wi-Fi system. With the information gathered from the OBU, it is feasible to warn vehicles and pedestrians of possible and imminent collisions. Besides, the tablet can keep the driver updated with critical supporting information.

The functional software architecture, as shown in Figure 2, can be composed of three parts. Regarding the OBU, it is fundamental to focus more attention on the implementation of the dedicated short-range communications (DSRC) and the coexistence between Wi-Fi and IEEE 802.11p. As for the smartphone, the goal is to provide reliable and efficient pedestrian warnings. In this sense, the notice of a possible collision arriving in the most appropriate mode can be provided, taking into account the detection of the specific state of the smartphone. The collision warning can also be provided to the tablet that is inside the vehicle. Notwithstanding, this warning will not prevent the driver from taking advantage of additional dedicated driving assistance information.

Regarding the coexistence between Wi-Fi and IEEE 802.11p, both protocols have different MAC80211 mechanisms. In most cases, it is difficult to run the IEEE 802.11p and Wi-Fi simultaneously. For instance, IEEE 802.11p [19] defines a way to exchange data without the need to establish a basic services set (BSS), while Wi-Fi protocols must connect to the BSS and wait for the association and authentication procedures before being able to exchange data. Therefore, the problem to be solved is to make it possible for these two different protocols to share MAC80211, which is one of the leading locks in the Linux wireless subsystem [33]. A flag serves to distinguish the IEEE 802.11p and Wi-Fi protocols in the MAC80211. The wireless network adapter is examined to identify the different protocols. In particular, in the struct net-device, there is a flag called ht-supported, which is used to define if a wireless network adapter can support a particular protocol, and, based on this, the protocol it can work with is established. It is necessary to note that the V2P communication for safety systems based on Wi-Fi could have some limitations:

- The battery level of mobile devices;
- some factors can affect the data transmission quality, such as obstacles, packets lost rate, and transmission delay;
- the accuracy or precision of the positioning device could not be satisfied with the requirement of the safety system.

Several systems are in a developing and researching phase, which, in most cases, have not been employed in the real world. Although most of the systems have validated the Wi-Fi reliability, the test carried out under particular conditions cannot be on behalf of the complex conditions in the real world. It is essential to examine more in detail these systems, like the one proposed in this paper in case of real implementation, with a particular focus on the reliability to improve the safety of the traffic situation. Besides, the vehicular ad-hoc network employed in the V2P safety is almost a new research field.

A careful assessment of the existing pieces of literature is essential for the development of Wi-Fi-based fault-tolerant solutions.

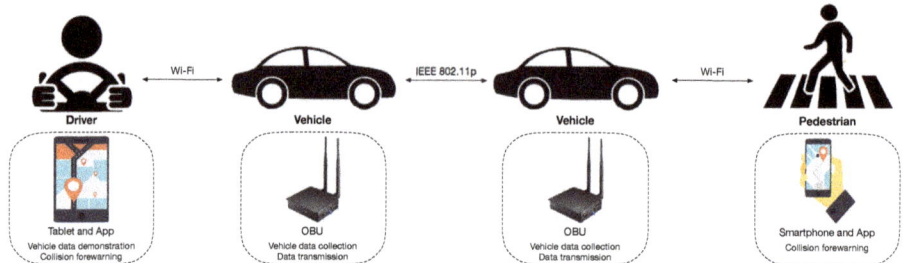

Figure 1. The proposed system architecture.

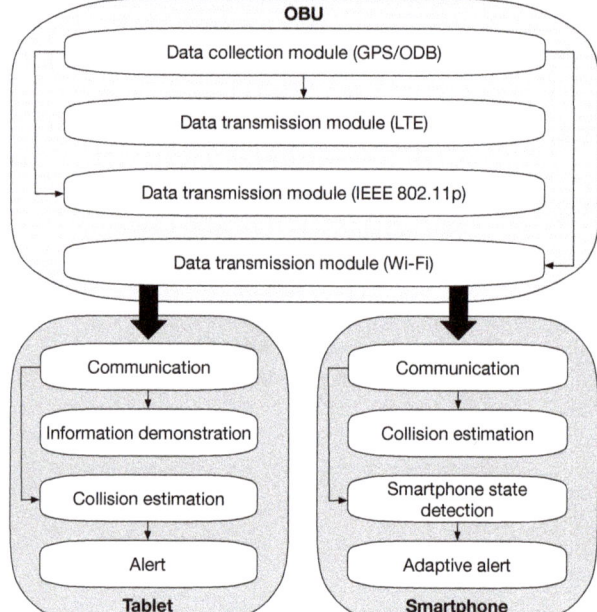

Figure 2. The proposed software architecture.

Before introducing the different characteristics of applications for the safety of drivers and pedestrians, it is advisable to perform a brief analysis regarding other distinctions of V2X communications. They are not presumed to have high latency toleration because safety-related applications require ultra-low latency [34]. For instance, pre-crash sensing alarms typically expect a cumulative latency of less than 50 ms [35]. It is necessary to note that V2V communications based on IEEE 802.11p, in most circumstances, must be conducted in non-line-of-sight conditions, while those that are cellular-based operate more trustworthily in transmission reliability, but produce more significant latency than IEEE 802.11p. Moreover, Wi-Fi-based V2P communication is useful when the transmission range is less than 150 m. Considering that these protocols are taken into account in the approach proposed in this paper, in a possible concrete implementation, these metrics should be recognized. Another characteristic to evaluate is related to communication security [36,37]. Data exchange between vehicles and other elements can be complicated as numerous data formats are transferred within various networks. One crucial concern is whether the data obtained from the V2X networks can be trusted, primarily if it is practiced for safety-related applications. For instance, an intruder could alter the transmitting operation to broadcast incorrect data deliberately. Further, data transferred in the V2X network might be modified by an attacker to mislead vehicles, which may lead to traffic accidents. Proper security schemes can be adopted, but extra latency may be introduced [38].

3. Concepts for Safety Applications

3.1. Vehicle Safety Application

Since the OBU does not have a user interface (UI) to interact with the driver, the vehicle security application is developed on a tablet and allows driving information to be displayed and collision warnings to be provided. After connecting to the OBU's Wi-Fi, the tablet can get the driving information of the surrounding vehicles. The GPS module installed in the OBU can provide location and speed information. The OBU allows for controlling several essential vehicle parameters, such as the speed or current fuel consumption, through the OBD-RS232 interpreter installed in the vehicle. It is useful to note that the measurements relating to the speed supplied by the OBD are more precise than those provided by the GPS.

As mentioned above, the OBU can obtain driving information on surrounding vehicles through the use of the IEEE 802.11p. As a consequence, the latter can also be used to estimate an incoming collision. In general, after receiving a package via IEEE 802.11p, the OBU can transfer it through Wi-Fi to the tablets and smartphones. With all this information obtained via Wi-Fi, thanks to the vehicular ad-hoc network (VANET) [39–44] application, it is possible to help the driver in monitoring vehicle status and threats from surrounding vehicles [45]. When the tablet estimates an incoming collision, it will inform the driver with a warning on the screen and a voice alert. The detailed information screen can display some parameters related to the surrounding vehicles, such as the position (longitude and latitude), driving speed, direction, and acceleration. After the tablet has received information on surrounding vehicles, it will notify the driver about the possibility of these vehicles generating a collision, thus ensuring more excellent safety for motorists. Moreover, attention appears to be focused mainly near road intersections, areas typically more characterized by possible clashes between motor vehicles [46].

Finally, when the tablet receives messages from the vehicle on which it is installed and other data by the surrounding vehicles, it can translate the latitudes and longitudes of the rectangular coordinate system of the Gauss–Krueger plane. Then, it estimates the driving vector of each vehicle and finds potential vehicles that cross with the vehicle concerned. If the carriers have an intersection nearby, they will calculate the time of the current vehicle and the vehicle with which it could crash. If these two times are below the threshold of the safety reaction time, then these two vehicles will have the risk of crashing each other.

3.2. Pedestrian Safety Application

Parallel to the application developed for the safety of motor vehicles, a smartphone security application can also be developed, useful for safeguarding distracted pedestrians. With messages received from the OBU via Wi-Fi [47], the application can provide the alert in the most appropriate mode based on the state of the smartphone. Concerning smartphones, in recent years, there has been a significant growth regarding their performance. Even mid or low range devices achieve remarkable performance, which until a few years ago was unthinkable for smartphones of these categories. This goal has been made possible thanks to an implicit standardization of the components to be used in terms of CPU, memory, and wireless protocols, such as cellular networks, Bluetooth, and Wi-Fi. The manufacturers of the most popular operating systems (Apple—iOS; Google—Android) now differentiate their products only for very small functionality, mostly focused on the use of artificial intelligence or augmented reality. The functioning of the apps is almost the same in both operating systems. Consequently, the diversity of operating systems is not a problem. Some small differences in performance could be found between high-level and low-level smartphones due to the diversity of the used hardware. Anyhow, it is evident that, in the future, the minimum requirements regarding the hardware and version of the operating system could be standardized to make the system work appropriately without incurring possible problems.

A possible implementation of the application is described in the flowchart depicted in Figure 3. First, the smartphone must access the open hotspot from the OBU, so that it can receive driving information from surrounding vehicles. After receiving a new package from the OBU, the smartphone will perform the collision estimation to recognize if the pedestrian could be run over by the vehicle. The application reads the GPS data of the smartphone to obtain the pedestrian position and information concerning its movement. Thus, the latitudes and longitudes of the smartphone and vehicle can be translated into the rectangular Gauss–Krueger plane. Next, the moving vectors of the smartphone and the vehicle will be calculated to see if they will have a collision. If the two carriers collide, the time required for the pedestrian and the vehicle to reach the intersection will be measured [14]. If the absolute time difference is lower than the safety intersection time threshold and the pedestrian crossing time is lower than the safety reaction time threshold, it is assumed that the pedestrian is at risk of being run over. With a positive collision prediction signal, a pedestrian alert is provided.

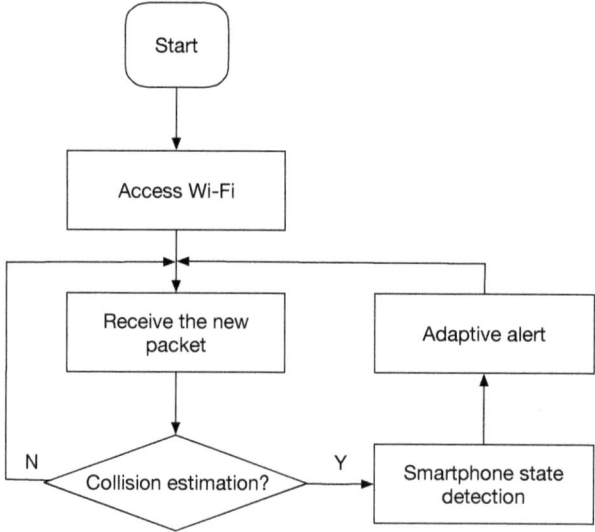

Figure 3. Application flowchart.

It has been already mentioned that pedestrians can use the smartphone in different ways. For each of these modes, the pedestrian will have to receive specific information, useful for grabbing attention during the specific activity performed. For instance, the specific examples can be classified into four categories: Reading an e-book, listening to music, watching and listening to a movie, inactive state. This classification is based on which specific part of the smartphone the user is focusing on. As a result, attention can be given to the screen rather than to the audio of the smartphone. The choice of the most suitable warning mode for the specific state of use is important to attract the pedestrian's attention. For example, a piece of sound information must be provided to capture the attention of the pedestrian. In each specific case, therefore, the most suitable warning mode corresponding to the state of use of the smartphone will be chosen. By analyzing the individual specific cases, it is possible to conclude that:

- In the screen-oriented state, the warning can consist of a message visible on the screen, i.e., the warning message is printed on the screen;

- a voice alert can be provided in the voice-oriented state. For instance, when the vehicle that could cause a collision with the pedestrian threat comes from the left side, the pedestrian will hear a distinctive voice that could state, "Collision forewarning! Please pay attention to the car on the left side";
- in the voice screen state, pedestrians can receive both the voice alert and the message on the screen simultaneously.

4. Conclusions

For several years, car manufacturers and government agencies have been looking at approaches to enhance road safety and effectively manage traffic flows. With the progress of wireless communication systems, the vision of cars that communicate with each other (V2V) and with systems placed at the edges of roads (V2I), this is becoming a reality. These wireless communication systems for vehicle-to-everything communications (V2X) are collectively known as intelligent transport systems (ITS). The numerous sensors that have been routinely installed on vehicles already detect information on distances to surrounding objects and their speeds. Anyhow, it is well known that the sharing of information from the vehicle to everything around it could allow acknowledging of surrounding situations so that the car driver can practice appropriate countermeasures to avoid accidents or get into the smoother traffic flow.

In this paper, the concepts for the development of a solution based on V2X communications aimed at vehicle and pedestrian safety have been presented. Regarding a potential system architecture for the development of a real system concerning the safety of vehicles and pedestrians, the drafting of guidelines that could be followed in new applications has been suggested. The primary purpose of this work has not been the introduction of a solution that has already been completed and experimented. As a result, performance evaluation has not been carried out. On the contrary, this paper has aimed at the introduction of the primary insights that a potential application for driver and pedestrian safety should support. To this end, a viable hardware and software architecture that could be employed in vehicles and mobile devices has been presented.

Further effort must be made to address some challenges related to this research field. For instance, the communication systems between vehicles require a stable wireless connection to maximize the transmission capacity of information in order to increase the safety of passengers. The receivers must be able to detect signals even in the worst conditions, e.g., low signal-to-noise ratio, poor modulation quality, and propagation conditions plagued by considerable fading, to ensure an excellent wireless connection. Furthermore, the rapid evolution of systems and components increases the complexity of the design. At the same time, the continuously updated regulations require a more significant number of tests and reliable and accurate results. Test budgets and methodologies fail to keep pace with the speed of innovation, putting pressure on test teams to reach results faster with fewer resources. The test must be able to respond to different types of requirements due to the need to create increasingly complex test systems, and be able to manage different types of measurements and, at the same time, the exponential growth of the collected data. This circumstance is not a problem only for the automotive industry. The same happens in similar sectors, such as heavy vehicles, aerospace, industrial machinery, and research. As a consequence, specific tools are required to optimize the workflow without sacrificing the flexibility to address these challenges.

Some future works concerning V2P applications could focus on mobile-accessible pedestrian signal systems. An application that enables an automatic call from the smartphone of the pedestrian who is blind or has low-vision to the traffic signal could be developed. Further, drivers endeavoring to perform a turn could be warned of the proximity of a pedestrian at the crosswalk. Other research could be performed warning of pedestrians at signalized crosswalks, for instance, developing an application that alerts transportation bus drivers when pedestrians, within the crosswalk of a signalized intersection, are in the expected path of the bus. Other efforts could be performed in investigating systems that can automatically brake a vehicle to avoid hitting a pedestrian, collectively referred to as pedestrian crash avoidance and mitigation systems. The large-scale dissemination of these methodologies could potentially reduce a considerable percentage of pedestrian crashes.

Author Contributions: The authors contributed equally to this work. All authors (G.P., F.A. and A.S.) have read and agreed to the published version of the manuscript.

Funding: This research received no external funding.

Acknowledgments: This work was related to the D.D. 407 of 27 February 2018 "AIM—Attrazione e Mobilità Internazionale" issued by the Italian Ministry of Education, University, and Research in implementation of Action I.2 "Mobilità dei Ricercatori" Asse I— PON R&I 2014–2020, taking into account the written amendment procedure of the PON R&I 2014–2020, pursuant to articles 30 and 90 of Regulation (EU) 1303/2013 started on 21 February 2018 as well as the relevant implementation regulations.

Conflicts of Interest: The authors declare no conflict of interest.

References

1. Neto, V.; Medeiros, D.; Campista, M. Analysis of mobile user behavior in vehicular social networks. In Proceedings of the 2016 7th International Conference on the Network of the Future (NOF), Buzios, Brazil, 16–18 November 2016; pp. 1–5. [CrossRef]
2. Deb, S.; Rahman, M.M.; Strawderman, L.J.; Garrison, T.M. Pedestrians' Receptivity Toward Fully Automated Vehicles: Research Review and Roadmap for Future Research. *IEEE Trans. Hum.-Mach. Syst.* **2018**, *48*, 279–290. [CrossRef]
3. Uchida, N.; Takeuchi, S.; Ishida, T.; Shibata, Y. Mobile traffic accident prevention system based on chronological changes of wireless signals and sensors. *J. Wirel. Mob. Netw. Ubiquitous Comput. Dependable Appl.* **2017**, *8*, 57–66. [CrossRef]
4. Giliberto, M.; Arena, F.; Pau, G. A fuzzy-based solution for optimized management of energy consumption in e-bikes. *J. Wirel. Mob. Netw. Ubiquitous Comput. Dependable Appl.* **2019**, *10*, 45–64. [CrossRef]
5. Larue, G.; Watling, C.; Black, A.; Wood, J.; Khakzar, M. Pedestrians distracted by their smartphone: Are in-ground flashing lights catching their attention? A laboratory study. *Accid. Anal. Prev.* **2020**, *134*. [CrossRef]
6. Song, F.; Zhu, M.; Zhou, Y.; You, I.; Zhang, H. Smart Collaborative Tracking for Ubiquitous Power IoT in Edge-Cloud Interplay Domain. *IEEE Internet Things J.* **2019**. [CrossRef]
7. Rangesh, A.; Trivedi, M.M. When Vehicles See Pedestrians With Phones: A Multicue Framework for Recognizing Phone-Based Activities of Pedestrians. *IEEE Trans. Intell. Veh.* **2018**, *3*, 218–227. [CrossRef]
8. Toppan, A.; Bazzi, A.; Toppan, P.; Masini, B.; Andrisano, O. Architecture of a simulation platform for the smart navigation service investigation. In Proceedings of the 2010 IEEE 6th International Conference on Wireless and Mobile Computing, Networking and Communications, Niagara Falls, ON, Canada, 11–13 October 2010; pp. 548–554. [CrossRef]
9. Pau, G.; Campisi, T.; Canale, A.; Severino, A.; Collotta, M.; Tesoriere, G. Smart pedestrian crossing management at traffic light junctions through a fuzzy-based approach. *Future Internet* **2018**, *10*. [CrossRef]
10. Tesoriere, G.; Campisi, T.; Canale, A.; Severino, A. The effects of urban traffic noise on children at kindergarten and primary school: A case study in Enna. In *AIP Conference Proceedings*; AIP Publishing: Melville, NY, USA, 2018; Volume 2040, p. 140005. [CrossRef]
11. Ticali, D.; Denaro, M.; Barracco, A.; Guerrieri, M. Piezoelectric energy harvesting from raised crosswalk devices. In *AIP Conference Proceedings*; AIP Publishing: Melville, NY, USA, 2015; Volume 1648. [CrossRef]
12. Arena, F.; Ticali, D. The development of autonomous driving vehicles in tomorrow's smart cities mobility. In *AIP Conference Proceedings*; AIP Publishing: Melville, NY, USA, 2018; Volume 2040. [CrossRef]
13. Xia, S.; de Godoy Peixoto, D.; Islam, B.; Islam, M.T.; Nirjon, S.; Kinget, P.R.; Jiang, X. Improving Pedestrian Safety in Cities Using Intelligent Wearable Systems. *IEEE Internet Things J.* **2019**, *6*, 7497–7514. [CrossRef]
14. Ai, Z.; Liu, Y.; Song, F.; Zhang, H. A Smart Collaborative Charging Algorithm for Mobile Power Distribution in 5G Networks. *IEEE Access* **2018**, *6*, 28668–28679. [CrossRef]
15. Song, F.; Ai, Z.; Zhou, Y.; You, I.; Choo, R.; Zhang, H. Smart Collaborative Automation for Receive Buffer Control in Multipath Industrial Networks. *IEEE Trans. Ind. Inform.* **2019**. [CrossRef]
16. Arena, F.; Pau, G. An overview of vehicular communications. *Future Internet* **2019**, *11*. [CrossRef]

17. Silva, C.; Silva, L.; Santos, L.; Sarubbi, J.; Pitsillides, A. Broadening understanding on managing the communication infrastructure in vehicular networks: Customizing the coverage using the Delta Network. *Future Internet* **2018**, *11*. [CrossRef]
18. Sheu, S.; Hsie, P.; Cheng, Y.; Wu, J. Coordinated scheduling scheme for IEEE 802.11p wireless vehicular networks using centralized scheduling of IEEE 802.16e wireless metropolitan area networks. In Proceedings of the 2011 7th International Wireless Communications and Mobile Computing Conference, Istanbul, Turkey, 4–8 July 2011; pp. 772–776. [CrossRef]
19. *IEEE Std 802.11p-2010 (Amendment to IEEE Std 802.11-2007 as Amended by IEEE Std 802.11k-2008, IEEE Std 802.11r-2008, IEEE Std 802.11y-2008, IEEE Std 802.11n-2009, and IEEE Std 802.11w-2009)*; IEEE: Piscataway, NJ, USA, 2010; pp. 1–51. [CrossRef]
20. Hancke, G.P.; Silva Bde, C.; Hancke, G.P., Jr. The Role of Advanced Sensing in Smart Cities. *Sensors* **2013**, *13*, 393–425. [CrossRef] [PubMed]
21. Masini, B.; Zuliani, L.; Andrisano, O. On the effectiveness of a GPRS based intelligent transportation system in a realistic scenario. In Proceedings of the 2006 IEEE 63rd Vehicular Technology Conference, Melbourne, Australia, 7–10 May 2006; Volume 6, pp. 2997–3001.
22. Tesoriere, G.; Campisi, T.; Canale, A.; Severino, A.; Arena, F. Modelling and simulation of passenger flow distribution at terminal of Catania airport. In *AIP Conference Proceedings*; AIP Publishing: Melville, NY, USA, 2018; Volume 2040. [CrossRef]
23. Cecchini, G.; Bazzi, A.; Masini, B.; Zanella, A. Performance comparison between IEEE 802.11p and LTE-V2V in-coverage and out-of-coverage for cooperative awareness. In Proceedings of the 2017 IEEE Vehicular Networking Conference (VNC), Torino, Italy, 27–29 November 2017; pp. 109–114.
24. Ticali, D.; Acampa, G.; Denaro, M. Renewable energy efficiency by railway transit. Case study on Rebaudengo railway tunnel in Turin. In *AIP Conference Proceedings*; AIP Publishing: Melville, NY, USA, 2018; Volume 2040. [CrossRef]
25. Bazzi, A.; Zanella, A.; Masini, B. A distributed virtual traffic light algorithm exploiting short range V2V Communications. *Ad Hoc Netw.* **2016**, *49*, 42–57. [CrossRef]
26. Balador, A.; Uhlemann, E.; Calafate, C.; Cano, J.C. Supporting beacon and event-driven messages in vehicular platoons through token-based strategies. *Sensors* **2018**, *18*. [CrossRef]
27. Zhenyu, L.; Lin, P.; Konglin, Z.; Lin, Z. Design and evaluation of V2X communication system for vehicle and pedestrian safety. *J. China Univ. Posts Telecommun.* **2015**, *22*, 18–26. [CrossRef]
28. Capallera, M.; Carrino, S.; De Salis, E.; Khaled, O.; Meteier, Q.; Angelini, L.; Mugellini, E. Secondary task and situation awareness, a mobile application for conditionally automated vehicles. In Proceedings of the 11th International Conference on Automotive User Interfaces and Interactive Vehicular Applications: Adjunct Proceedings, Utrecht, The Netherlands, 21–25 September 2019; pp. 86–92.
29. Veena, S.; Ramyadevi, K.; Elavarasi, K.; Preetha, M. Smart car automated system to assist the driverin detecting the problem and providing the solution. *Int. J. Innov. Technol. Explor. Eng.* **2019**, *8*, 727–731. [CrossRef]
30. Nguyen, Q.H.; Morold, M.; David, K.; Dressler, F. Car-to-Pedestrian communication with MEC-support for adaptive safety of Vulnerable Road Users. *Comput. Commun.* **2020**, *150*, 83–93. [CrossRef]
31. Song, F.; Zhou, Y.; Wang, Y.; Zhao, T.; You, I.; Zhang, H. Smart Collaborative Distribution for Privacy Enhancement in Moving Target Defense. *Inform. Sci.* **2019**, *479*, 593–606. [CrossRef]
32. Wang, S.; Yuan, J.; Li, X.; Qian, Z.; Arena, F.; You, I. Active Data Replica Recovery for Quality-Assurance Big Data Analysis in IC-IoT. *IEEE Access* **2019**, *7*, 106997–107005. [CrossRef]
33. Linux Wireless Wiki. Available online https://wireless.wiki.kernel.org (accessed on 31 October 2019).
34. Qiu, H.; Qiu, M.; Lu, Z.; Memmi, G. An efficient key distribution system for data fusion in V2X heterogeneous networks. *Inf. Fusion* **2019**, *50*, 212–220.[CrossRef]
35. Filippi, A.; Moerman, K.; Daalderop, G.; Alexandera, P.D.; Schober, F.; Pfliegl, W. Ready to roll: why 802.11p beats LTE and 5G for V2X, A White Paper. Available online: https://assets.new.siemens.com/siemens/assets/public.1510309207.ab5935c545ee430a94910921b8ec75f3c17bab6c.its-g5-ready-to-roll-en.pdf (accessed on 13 November 2019).
36. Khodaei, M.; Noroozi, H.; Papadimitratos, P. Scaling Pseudonymous Authentication for Large Mobile Systems. In Proceedings of the WiSec '19 Proceedings of the 12th Conference on Security and Privacy in Wireless and Mobile Networks, Miami, FL, USA, 15–17 May 2019; pp. 174–184. [CrossRef]

37. Qiu, H.; Kapusta, K.; Lu, Z.; Qiu, M.; Memmi, G. All-Or-Nothing data protection for ubiquitous communication: Challenges and perspectives. *Inf. Sci.* **2019**, *502*, 434–445. [CrossRef]
38. Qiu, H.; Qiu, M.; Lu, R. Secure V2X Communication Network based on Intelligent PKI and Edge Computing. *IEEE Netw.* **2019**, 1–7. [CrossRef]
39. Santamaria, A.F.; Tropea, M.; Fazio, P.; De Rango, F. Managing Emergency Situations in VANET Through Heterogeneous Technologies Cooperation. *Sensors* **2018**, *18*, 1461. [CrossRef] [PubMed]
40. Medeiros, D.S.; Hernandez, D.A.; Campista, M.E.M.; Aloysio de Castro, P.P. Impact of relative speed on node vicinity dynamics in VANETs. *Wirel. Netw.* **2019**, *25*, 1895–1912. [CrossRef]
41. Bazzi, A.; Zanella, A.; Masini, B. An OFDMA-Based MAC Protocol for Next-Generation VANETs. *IEEE Trans. Veh. Technol.* **2015**, *64*, 4088–4100. [CrossRef]
42. Nkenyereye, L.; Tama, B.; Park, Y.; Rhee, K. A fine-grained privacy preserving protocol over attribute based access control for VANETs. *J. Wirel. Mob. Netw. Ubiquitous Comput. Dependable Appl.* **2015**, *6*, 98–112.
43. Sarubbi, J.; Silva, T.; Martins, F.; Wanner, E.; Silva, C. Allocating Roadside Units in VANETs Using a Variable Neighborhood Search Strategy. In Proceedings of the 2017 IEEE 85th Vehicular Technology Conference (VTC Spring), Sydney, Australia, 4–7 June 2017; pp. 1–5. [CrossRef]
44. Eziama, E.; Tepe, K.; Balador, A.; Nwizege, K.; Jaimes, L. Malicious Node Detection in Vehicular Ad-Hoc Network Using Machine Learning and Deep Learning. In Proceedings of the 2018 IEEE Globecom Workshops (GC Wkshps), Abu Dhabi, UAE, 9–13 December 2018; pp. 1–6. [CrossRef]
45. Brahmi, H.I.; Djahel, S.; Murphy, J. Improving emergency messages transmission delay in road monitoring based WSNs. In Proceedings of the 6th Joint IFIP Wireless and Mobile Networking Conference (WMNC), Dubai, UAE, 23–25 April 2013; pp. 1–8. [CrossRef]
46. Chen, C.; Chen, P.; Chen, W. A Novel Emergency Vehicle Dispatching System. In Proceedings of the 2013 IEEE 77th Vehicular Technology Conference (VTC Spring), Dresden, Germany, 2–5 June 2013; pp. 1–5. [CrossRef]
47. Vaculík, M.; Franeková, M.; Vestenický, P.; Vestenický, M. On-board Unit and its Possibilities of Communications on Safety and Security Principles. *Adv. Electr. Electron. Eng.* **2008**, *7*, 235–238.

© 2019 by the authors. Licensee MDPI, Basel, Switzerland. This article is an open access article distributed under the terms and conditions of the Creative Commons Attribution (CC BY) license (http://creativecommons.org/licenses/by/4.0/).

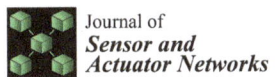

Article

Scalable Edge Computing Deployment for Reliable Service Provisioning in Vehicular Networks

Federico Tonini, Bahare M. Khorsandi, Elisabetta Amato and Carla Raffaelli *

DEI—University of Bologna, Viale del Risorgimento, 2, 40136 , Bologna, Italy; f.tonini@unibo.it (F.T.); bahare.masood@unibo.it (B.M.K.); e.amato@unibo.it (E.A.)
* Correspondence: carla.raffaelli@unibo.it; Tel.: +39-051-209-3058

Received: 2 September 2019 ; Accepted: 29 September 2019; Published: 2 October 2019

Abstract: The global connected cars market is growing rapidly. Novel services will be offered to vehicles, many of them requiring low-latency and high-reliability networking solutions. The Cloud Radio Access Network (C-RAN) paradigm, thanks to the centralization and virtualization of baseband functions, offers numerous advantages in terms of costs and mobile radio performance. C-RAN can be deployed in conjunction with a Multi-access Edge Computing (MEC) infrastructure, bringing services close to vehicles supporting time-critical applications. However, a massive deployment of computational resources at the edge may be costly, especially when reliability requirements demand deployment of redundant resources. In this context, cost optimization based on integer linear programming may result in being too complex when the number of involved nodes is more than a few tens. This paper proposes a scalable approach for C-RAN and MEC computational resource deployment with protection against single-edge node failure. A two-step hybrid model is proposed to alleviate the computational complexity of the integer programming model when edge computing resources are located in physical nodes. Results show the effectiveness of the proposed hybrid strategy in finding optimal or near-optimal solutions with different network sizes and with affordable computational effort.

Keywords: vehicular networks; 5G; C-RAN; resource allocation; edge computing; optimization

1. Introduction

Connected vehicles can provide a large set of services for smarter and safer mobility. As an example, a problem highly felt worldwide is road safety [1] where vehicular networks can help in providing prompt information to drivers and alerting possible dangerous situations by allowing vehicles to communicate with each other. It is possible to distinguish between short-range direct communications and long-range network communications [2,3]. Different communications need different network requirements, such as low latency, high computational capacity, and high reliability, depending on the application [4].

To provide the aforementioned services, 5G networks can be used to carry data to/from vehicles and road infrastructure. Centralized cloud-based Radio Access Networks (C-RANs) represent an effective solution to design high-capacity radio access in 4G and 5G networks and to support challenging use cases [5], such as the ones of vehicular networks. C-RAN introduces unprecedented flexibility by efficient application of Network Function Virtualization (NFV) [6] jointly with Software Defined Networking (SDN) [5,7,8]. SDN can, in fact, provide suitable control and management support to optimally locate virtualized network functionalities to intelligent nodes in the cost and power efficiency perspectives. This is of particular importance when considering highly dynamic and performance-constrained contexts as happens in 5G networking. In addition, to ensure timely network adaptation to user needs, SDN control and management must cope with a potentially high number of network elements and, consequently, the design of control algorithms calls for highly scalable approaches. Virtualized baseband functionalities are suitably located and centralized in the nodes of the optical transport network implementing a C-RAN for enhanced functionality and cost-optimization purposes [7]. The nodes hosting these pooled virtual baseband units (BBUs) are called BBU hotels. BBU hotels can be provided with an additional computational capacity to perform time-sensitive operations required by low-latency services, as per the Multi-access Edge Computing (MEC) paradigm [9]. MEC, by providing 5G with processing resources at the network edge, allows to achieve stringent application requirements. However, widespread deployment of these nodes may be costly; therefore, intelligent nodes hosting BBU hotels and edge computing resources need to be identified in relation to latency and processing constraints. Moreover, the problem of BBU hotel placement in C-RAN has been shown to be NP-hard [10], requiring novel strategies to make optimal approaches more scalable.

This paper proposes an Integer Linear Program (ILP) to solve the joint deployment problem of baseband processing and edge computing with reliability against single-node failure in C-RAN. The main objective of this strategy is to minimize the nodes in which processing capabilities must be installed while ensuring latency and optical link (i.e., maximum wavelengths over fibers) constraints are not violated. To overcome the computational complexity of classical optimization approaches, a hybrid (based on both heuristic and ILP) deployment strategy is also proposed. The algorithm performs a first phase in which the initial set of nodes candidate to host baseband and edge computing functions is reduced and a suboptimal solution is provided. Then, a second phase is executed for optimization purposes. The latter approach is shown to provide results close to optimal ones while considerably reducing computational time.

The paper is organized as follows. In Section 2, related works in the context are introduced. Section 3 provides an overview of the reference C-RAN architecture and describes the deployment problem. In Section 4, the optimization problem is formulated, while Section 5 presents the hybrid model. In Section 6, the numerical results obtained in different scenarios are presented. Finally, Section 7 concludes the paper.

2. Background

In Reference [11], the Vehicular Edge Computing (VEC) architecture is analyzed. VEC is composed of three layers: users, MEC, and cloud. In the user layer, vehicles exchange information with each other or with road infrastructure through different protocols, such as the ones reported in References [12,13]. Mobile networks can also be used for vehicular communications to directly send/retrieve data to/from vehicles or road infrastructure [14]. Data are then carried to the MEC or cloud layers, where different services are located. The MEC layer is usually needed to provide low-latency services or to offload the network by means of content caching [14]. The cloud layer, instead, is located deeper in the network and offers extensive computational capacity for heavy data processing and non-time-sensitive applications.

The adoption of 5G new radio allows to use its efficient cell coordination and interference management mechanisms as well as novel discovery techniques to improve performance in dense scenarios. With 5G deployments, C-RAN will gradually take over conventional distributed networks in favor of more efficient centralized networks [15]. In C-RAN, antennas and Remote Radio Units (RRUs) are usually located at antenna sites, while BBUs are decoupled from RRUs and placed in central locations [5]. Part or all the baseband processing functions realizing the mobile network protocol stack are performed in the BBUs, where they can be virtualized over general-purpose hardware to decrease network costs [5]. Depending on the specific functions performed in the different units, traffic with different characteristics is transported over the so-called fronthaul links interconnecting them [16]. According to the adopted functional split, the nodes of the transport network can operate at different layers of the protocol stack [17]. Multiple splits can also be defined depending on different levels of baseband function centralization [18,19]. MEC will also play a fundamental role in 5G to meet low-latency service requirements [9,20]. Recently, different architectures have been proposed for edge computing [21]. Edge data centers can be co-located with 4G or 5G baseband processing functions to reduce the delay by bringing services closer to the users [9]. This is of particular relevance for vehicular networks that can take advantage of this to offer Ultra Reliable and Low-Latency Communication services (URLLC) over 5G networks. However, the cost for a large-scale reliable deployment of edge core and cloud resources must be taken into consideration and calls for cost-efficient deployment strategies.

In References [22,23], the authors present optimal and suboptimal strategies for placement of edge resources in 5G networks. In Reference [22], a framework to optimize the placement of primary and backup 5G user plane functions (UPFs) at the edge is provided. The proposed deployment strategies aim at configuring edge resources at a minimum cost while ensuring service demands are met. Results show the amount and use of required UPF for different scenarios and provide a complexity analysis and the execution time of different algorithms. However, the proposed model considers only backhaul links and do not account for the finite optical link capacity, which may affect the solution, especially when dealing with very high bitrate requirements of Fronthaul links. The model in Reference [23], instead, focuses on the number of edge nodes to be equipped with computational capacity, which is shown to increase with the number of base stations deployed in the area.

Deployment strategies for C-RAN have been proposed recently for Wavelenght Division Multiplexing (WDM) networks based on ILP and heuristic strategies [24,25]. Reliability aspects for C-RAN deployment are analyzed in detail in References [10,26,27]. In References [28], the authors propose a fog computing framework for C-RAN in vehicular networks. Simulation results show that low latency can be achieved with edge computing under different traffic conditions. However, no consideration has been made on the deployment of C-RAN processing functions in access/aggregation networks. In Reference [29], the authors propose an edge server placement for MEC in distributed RAN based on integer programming. The proposed approach is compared with different benchmarking strategies. Results show how the different strategies perform in terms of edge service access delay for different numbers of edge nodes using a realistic dataset. While the focus of this work is on distributed RAN, in C-RAN, specific constraints on latency and bandwidth requirements set by the baseband processing at the physical layer are needed. In addition, resiliency aspects are not considered in Reference [29]. The work in Reference [30] proposes an ILP and a heuristic for the deployment of cloud fog RAN. The authors conduct an extensive analysis of the trade-offs among the minimization of propagation latency and power consumption, but no mention of reliability aspects against failures is made. The work in Reference [26] is extended here with proper routing (i.e., not based on precomputed shortest path), and considerations on edge computing deployment for URLLC services in C-RAN are made. A novel heuristic strategy is also proposed to reduce the computational complexity of the optimization problem by properly reducing the set of possible locations for C-RAN and MEC infrastructure.

3. Architectural Solution and Problem Formulation

The reference C-RAN architecture has been introduced in previous works [7,31]. It consists of a hierarchical SDN control plane with a lower layer split into as many controllers as the different kinds of network domains to control, namely the radio network, the optical transport network, and the cloud network. An example of this architectural solution applied to vehicular scenarios is shown is Figure 1. The radio domain is composed of antennas and RRUs located at cell sites, and baseband processing functions that are performed over general-purpose hardware in edge nodes. The radio controller is in charge of controlling radio and baseband resources that are remotized following the C-RAN design concept. The optical transport network consists of a set of intelligent nodes interconnected by Dense Wavelength Division Multiplexing (DWDM) optical links to support high-capacity fronthaul in C-RAN. For example, to support heavy and constant fronthaul traffic generated by the Common Public Radio Interface (CPRI) split [19] (referred to as Option 8 in Reference [18]), dedicated wavelengths are usually required. Nodes of the transport network, referred to here as edge nodes, are equipped with processing capabilities to perform MEC functionalities and are managed by the cloud controller. Each controller interacts with the SDN orchestrator to provide information for interworking control and management functions through different domains. The orchestrator is in charge of accommodating new service requests by suitably allocating required resources across the different domains. The orchestrator applies suitable algorithms to properly select the nodes in which the BBU functionalities and services are executed, depending on service and physical network constraints.

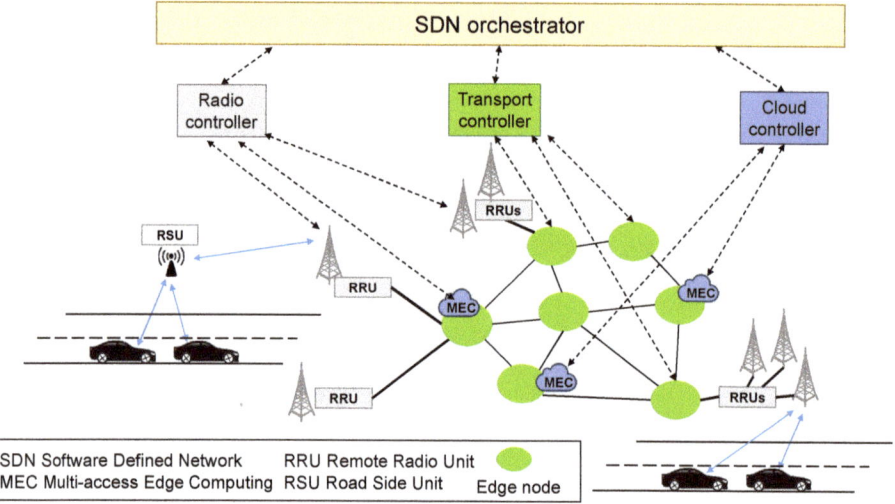

Figure 1. Softward Defined Networking (SDN)-controlled Cloud Radio Access Network (C-RAN) architecture for vehicular communications.

C-RAN architecture can be used as an enabler for vehicular communications providing network assistance and commercial services, as depicted in Figure 1. Vehicles communicate directly with the mobile network or with Road Side Units (RSUs), that send collected data through the mobile network. Data concerning low-latency applications can be elaborated directly in the edge nodes, thanks to the computational resources offered by the MEC. Computational resources in edge nodes can be used for (i) virtual baseband processing; (ii) virtual mobile core network functions; and (iii) edge application services [32]. Non-time-sensitive data can be delivered to applications performed in remote locations (not reported in the figure). The traffic destined to remote cloud resources is user dependent and requires lower bandwidth with respect to fronthaul requirements [16] and is out of the scope of this paper. In this work, we propose to co-locate, within the same edge node, cloud and BBU processing functions. An edge node is considered to be active when it hosts physical or virtual functions, either for BBU processing or edge core/cloud services.

To provide a reliable C-RAN against single node failures, a 1 + 1 protection solution is desirable to avoid temporary service outages due to resource restoration. Primary and backup path resources must be allocated to provide resiliency against hardware failures. This work considers single active edge node failures (i.e., a failure of all servers placed in an active edge node). The formulation of the joint BBU hotel and edge cloud processing location problem with resiliency is as follows:

- **Given** a set of RRUs to be connected to active edge nodes, a set of edge nodes (candidates to host BBU and edge processing resources), and a set of links connecting edge nodes.
- **Find** active edge nodes and suitable optical resource assignment such that (i) the number of active nodes and (ii) total wavelengths are minimized.

- **Ensure** that each RRU is connected to two active edge nodes (one for primary and one for backup purposes) and that the maximum available wavelengths per link and maximum allowed distance to provide target service are not exceeded.

4. ILP-Based Optimization

This section proposes an ILP formulation of the problem. This algorithm is expected to be executed by the orchestrator, which is assumed to have complete knowledge of the underlying network topology and available resources to provide the placement. The notation used in the algorithm is reported in Table 1. The set of nodes in the network, the candidate to host BBU and edge processing functions, is denoted as N, while the number of sources (RRUs) physically connected to node $s \in N$ is denoted as R_s. The connectivity among them is modeled by the C binary matrix. C has one row and one column for each node, and an element is equal to 1 if the two nodes are directly connected by a link, 0 otherwise. Binary variables p_{sd}^H and b_{sd}^H are equal to 1 if node $d \in N$ is the node processing data from RRUs located at node s for primary or backup, respectively. The binary variable h_d is equal to 1 if edge node d is active, i.e., if it acts as a primary or a backup for one or more RRUs. $h_d = 1$ also means that at least one between p_{sd}^H and b_{sd}^H is equal to 1. To connect each RRU to the nodes performing processing functions, one wavelength is reserved along the path, due to the high requirements of physical layer processing functions. This is captured by binary variables w_{sdij}^p and w_{sdij}^b. The maximum available wavelengths over each link and the maximum allowed distance between RRUs and BBUs are indicated with M^W and M^H, respectively. In this formulation, edge processing functions are co-located with BBU processing to reduce the delay to a minimum and take advantage of the already active nodes, without requiring additional resources on fibers to reach farther facilities. For this reason, only M^H is considered, which is usually more stringent. If this is not the case, M^H could represent the service delay and be used as a more stringent delay requirement. In this work, all links are assumed to be equally long, so M^H is expressed in terms of hops.

The formulation is as follows.

Table 1. Notation for Integer Linear Program (ILP).

Parameter	Definition
N	set of edge nodes in the network, $|N| = n$.
R_s	number of sources (RRUs) directly connected to $s \in N$.
C	$n \times n$ matrix. $c_{ij} = 1$ if node i is directly connected to node j, 0 otherwise.
p_{sd}^H	binary variable, equal to 1 if edge node $d \in N$ acts as primary for RRUs at node (cell site) $s \in N$; 0 otherwise.
b_{sd}^H	binary variable, equal to 1 if edge node $d \in N$ acts as backup for RRUs at node (cell site) $s \in N$; 0 otherwise.
h_d	binary variable equal to 1 if edge node $d \in N$ is active, 0 otherwise.
w_{sdij}^p	binary variable, equal to 1 if the path to connect RRUs at node $s \in N$ and primary edge node $d \in N$ is using physical link $i-j$ $(i,j \in N)$; 0 otherwise.
w_{sdij}^b	binary variable, equal to 1 if the path to connect RRUs at node $s \in N$ and backup edge node $d \in N$ is using physical link $i-j$ $(i,j \in N)$; 0 otherwise.
M^W	max. available wavelengths in each link.
M^H	max. allowed distance between RRUs and edge nodes.
$\alpha, \beta \in \mathbb{N}$	tuning parameters for the objective function.
$L \in \mathbb{N}$	a large number (e.g., 10,000).

Objective Function

$$\text{Minimize } F = \alpha \cdot \sum_{d \in N} h_d + \beta \cdot \sum_{s \in N} \sum_{d \in N} \sum_{i \in N} \sum_{j \in N} w_{sdij}^p + w_{sdij}^b \qquad (1)$$

The multi-objective function in Equation (1) is composed of two members. The first term takes into account the activation cost of each node, while the second term accounts for the wavelengths required to connect RRUs to edge nodes, both primary and backup.

The problem is subject to the following constraints:

$$\sum_{d \in N} p_{sd}^H = 1, \quad \forall s \in N \tag{2}$$

$$\sum_{d \in N} b_{sd}^H = 1, \quad \forall s \in N \tag{3}$$

$$p_{sd}^H + b_{sd}^H \leq 1, \quad \forall s, d \in N \tag{4}$$

$$h_d \cdot L \geq \sum_{s \in N} p_{sd}^H + b_{sd}^H, \quad \forall d \in N \tag{5}$$

$$\sum_{s \in N} \sum_{d \in N} (w_{sdij}^p + w_{sdij}^b + w_{sdji}^p + w_{sdji}^b) \cdot R_s \leq M^W, \quad \forall i, j \in N \tag{6}$$

$$w_{sdij}^p \leq c_{ij}, \quad \forall s, d, i, j \in N \tag{7}$$

$$w_{sdij}^b \leq c_{ij}, \quad \forall s, d, i, j \in N \tag{8}$$

$$\sum_{i \in N} \sum_{j \in N} w_{sdij}^p \leq M^H, \quad \forall s, d \in N \tag{9}$$

$$\sum_{i \in N} \sum_{j \in N} w_{sdij}^b \leq M^H, \quad \forall s, d \in N \tag{10}$$

$$\sum_{i \in N} w_{sdij}^p - w_{sdji}^p = \begin{cases} p_{sd}^H & \text{if } j = s, s \neq d, \forall s, d, j \in N \\ -p_{sd}^H & \text{if } j = d, s \neq d, \forall s, d, j \in N \\ 0 & \text{otherwise} \end{cases} \tag{11}$$

$$\sum_{i \in N} w_{sdij}^b - w_{sdji}^b = \begin{cases} b_{sd}^H & \text{if } j = s, s \neq d, \forall s, d, j \in N \\ -b_{sd}^H & \text{if } j = d, s \neq d, \forall s, d, j \in N \\ 0 & \text{otherwise} \end{cases} \tag{12}$$

The constraints of Equations (2) and (3) ensure that there is only one primary and one backup edge node, respectively, for each RRU. The constraint of Equation (4) guarantees that primary and backup nodes are disjoint. The constraint of Equation (5) counts the number of active nodes (i.e., performing processing functions) in case they are acting either as a primary or a backup for any RRU. The constraint of Equation (6) limits the number of wavelengths over each link for both primary and backup in both directions (i.e., from i to j and j to i together). The constraints of Equations (7) and (8) ensure the feasibility of the connections so that a link between two nodes can be used if and only if it exists in the physical topology. The constraints of Equations (9) and (10) limit the maximum distance between RRUs and BBUs to M^H for primary and backup paths, respectively. Finally, Equations (11) and (12) are the flow conservation constraints for primary and backup paths, respectively. These constraints are needed to reserve the paths connecting RRUs to their primary and backup edge nodes. In this model, wavelength conversion is allowed in the network nodes.

5. Two-Phases Hybrid Approach

The hybrid approach proposed here is performed in two phases. In the first phase, a heuristic is proposed to provide a computationally simple but reliable C-RAN coverage by guaranteeing that each RRU has both a primary and a backup node and that minimum delay is achieved. The second phase is an optimization process, based on a modified version of the ILP proposed in Section 4, that aims at reducing the number of active nodes found in phase 1. The details of the hybrid algorithm are reported below.

Phase 1 is assumed to start from a C-RAN configuration where no edge node is active, i.e., BBU and edge functionalities have yet to be assigned to nodes. This has, anyway, no impact on the generality of the approach. In this phase, the edge node activation is performed within a 1-hop distance or, equivalently, RRUs can be connected only to the node itself or to a neighbor edge node. This implicitly assumes that there are enough resources on the links connecting neighbors and guarantees that delay constraints are always satisfied. It should be noted that, to solve the deployment problem, primary and backup nodes must be selected. Therefore, not satisfying the aforementioned condition on the link resources does not guarantee a solution to the problem.

In addition to the C matrix needed to model the physical links (see Table 1), two additional structures are introduced here:

- H matrix: This is an $n \times 2$ matrix, where each row represents a node of the network; the first column indicates which is the primary edge node chosen by the node on that row, while the second column indicates which is the backup node.
- W matrix: This is an $n \times n$ matrix which keeps track of the use of the links between nodes. In W, there is one row for each source edge node (where the RRUs are physically connected). W has one column for each edge node, that is, the possible locations for the edge server performing baseband and services for the specific RRUs. This matrix is needed to provide a feasible solution at the end of phase 1 but is not used in phase 2.

Algorithm 1 presents the pseudo-code of the algorithm executed by each node of the network during phase 1. In the beginning, the algorithm starts with empty H and W matrices (line 2). This algorithm executed in a sequence for each node until all nodes in the network have both primary and backup connections (condition in line 4). Then, node i checks some conditions for the primary and for the backup connection in order to find suitable edge nodes. If node i is already active (line 6), it can use itself as the primary edge node (line 7). Otherwise, node i must search among its neighbors to find an already active node (line 8) and, if it succeeds, makes the primary connection to the edge node j (line 9) and updates W matrix accordingly (line 10). The updating phase stores in the position i, j of the matrix the required wavelengths over link i–j. If no neighbor is active (line 11), node i activates itself and makes the primary connection to itself (lines 12 and 13).

After establishing the primary connection, node i executes a set of instructions to find the backup edge node. There are two possible situations. The first situation is when node i is already active and plays the primary role for the RRUs connected to itself or not active at all (line 16). In this case, node i either finds a directly connected neighbor node (j), which is already active and satisfies the distance restriction, and connects to it (lines 17–19) or chooses randomly one of the neighbors as a backup, defines the backup connection, and updates W matrix accordingly (lines 20–23). The other situation happens when node i is active (line 25). Node i can take advantage of this situation and makes the backup connection to the local edge node (lines 26 and 27). Phase 1 stops when all nodes in the network have both connections to primary and backup nodes.

The objective of the second phase is to minimize the number of active nodes. This is achieved by reassigning the RRU connections and shutting down active nodes by further centralizing BBU and edge processing functions within the distance constraints (M^H). This is achieved by adding the following set of constraints to the ILP model presented in Section 4:

Equation (13) forces the node candidates to be 0 (non-active) for all the nodes excluded by phase 1 (i.e., for all the nodes that have no RRU assigned to them, either for primary or backup purposes). The ILP is then solved with a reduced set of candidate nodes that always ensures the feasibility of the solution.

$$h_d = \begin{cases} 0 & \text{if } H_{d0} + H_{d1} = 0, \quad \forall d \in N \\ \{0,1\} & \text{otherwise} \end{cases} \qquad (13)$$

Algorithm 1 C-RAN reliable coverage (phase 1).

1: **Initialization:**
2: $H, W \leftarrow \emptyset$
3: **Begin:**
4: **while** exists node $i \in N$ s.t. $(H_{i0} = 0) \vee (H_{i1} = 0)$
5: //Primary connection assignment:
6:   **if** $h_j = 1$
7:     $H_{i0} = i$
8:   **else if** $\exists$ node j s.t. $c_{ij} = 1$ and $h_j = 1$
9:     $H_{i0} = j$
10:     update W
11:   **else**
12:     $h_i = 1$
13:     $H_{i0} = i$
14:   **end if**
15: //Backup connection assignment:
16:   **if** $(h_j = 1$ and $H_{i0} = i)$ or $(h_i = 0)$
17:     **if** $\exists$ node j s.t. $c_{ij} = 1$ and $h_j = 1$
18:       $H_{i1} = j$
19:       update W
20:     **else**
21:       **activate** random neighbor j $(h_j = 1)$
22:       $H_{i1} = j$
23:       update W
24:     **end if**
25:   **else**
26:     $h_i = 1$
27:     $H_{1i} = i$
28:   **end if**
29: **end while**
30: **End**

---

6. Numerical Results

Numerical results are obtained in different networks to evaluate the effectiveness of the ILP and hybrid solutions in terms of active edge nodes and of the centralization gain, G_C, that is the advantage related to centralizing BBU and cloud functionalities, expressed by the following formula:

$$G_C = \frac{|N| - \sum_{d \in N} h_d}{|N|} \qquad (14)$$

where $|N|$ and h_d have been defined in Table 1. Three sample networks, N_{38}, N_{20}, and N_{14}, consisting of 38, 20, and 14 nodes, respectively, are considered, as represented in Figure 2. Evaluations assume here that 10 RRUs are physically connected to each node to provide mobile network coverage and transmission capacity for vehicular network, and the adoption of CPRI (option 8 in Reference [18]). The proposed algorithms and evaluations can be extended to different numbers of RRUs, possibly unbalanced among edge nodes and suitably adapted to different functional split, which is left for future works. The commercial tool CPLEX [33] is used to run the ILP on a computer with 4 cores at 3.2 GHz and 8 GB of RAM. Tuning parameters α and β are set to a value of $\alpha \gg \beta$ so that the minimization of active edge nodes is prioritized, while the maximum number of wavelengths over each link M^W is set to 80.

In Figures 3–5, comparisons are reported between the hybrid and the ILP approaches by plotting the results in terms of the number of active edge nodes as a function of the allowed distance, expressed in hops. The cost of the hybrid solution depends on the node from which the heuristic procedure starts: the maximum and minimum costs in terms of total number of active nodes obtained are both reported in the plots. In addition, the results at the end of phase 1 of the hybrid strategy are also shown, as lines and denoted as H, to outline the effect of the optimization phase. These lines are constant because they do not depend on the distance, as they provide a solution within 1 hop distance. The costs obtained with the hybrid and ILP approaches decrease with the distance in all networks. The minimum value that can be achieved is 2 because one primary and one backup node must be always present to cope with single edge node failure. In case of tight distance constraints (e.g., 1 or 2 hops), data cannot be transported far in the network; thus, many edge nodes must be activated. When the distance constraint increases, farther nodes in the network can be reached and, consequently, the number of total active nodes decreases. From the figures, it can be seen also the influence of the starting node, represented by the difference between the maximum and the minimum costs. In the worst cases, only one additional node must be activated. In addition, the results of the hybrid are shown to be the same as the optimal ones in most of the cases. However, in very few cases, the hybrid approach cannot achieve optimal solutions due to the choices performed in phase 1, where some nodes are excluded by the pool of possible active nodes and cannot be activated in phase 2.

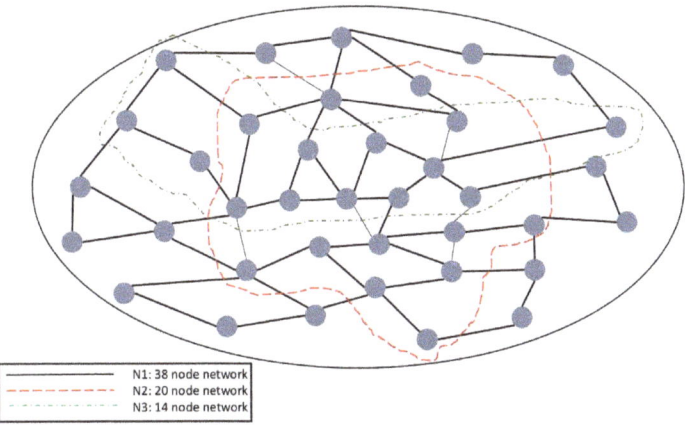

Figure 2. N_{38}, N_{20}, and N_{14} C-RAN topology for numerical evaluations.

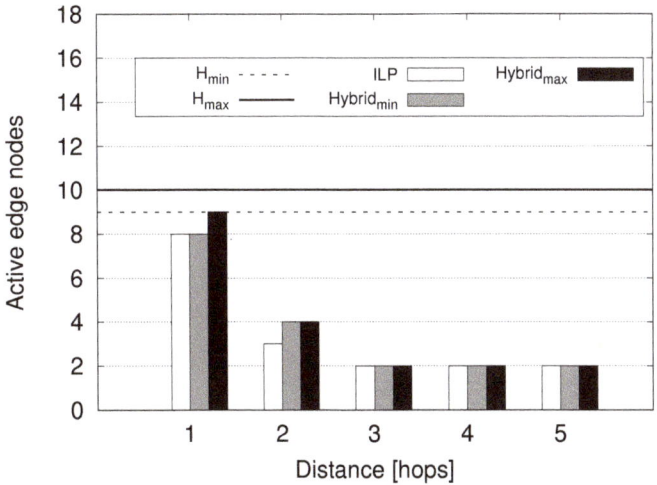

Figure 3. Total number of active edge nodes as a function of the allowed distance between RRUs and edge nodes for network N_{14}: Maximum and minimum costs of the hybrid results are reported after both phases.

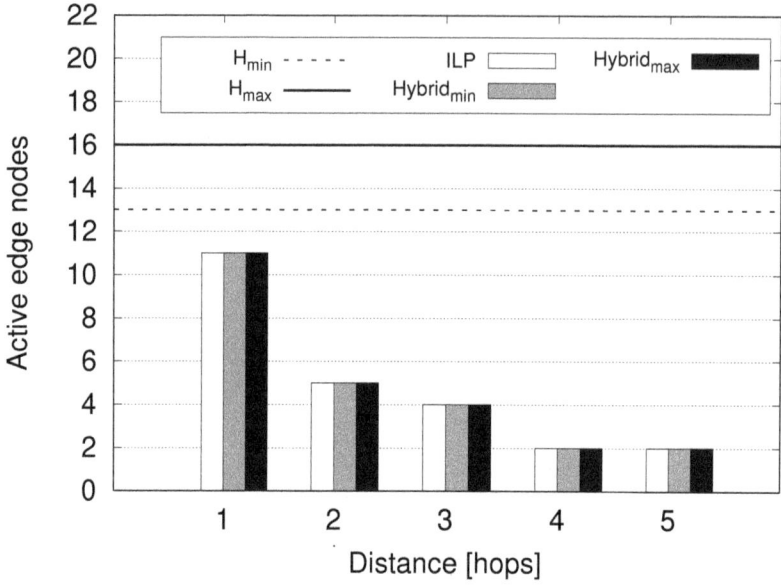

Figure 4. Total number of active edge nodes as a function of the allowed distance between RRUs and edge nodes for network N_{20}: Maximum and minimum costs of the hybrid results are reported after both phases.

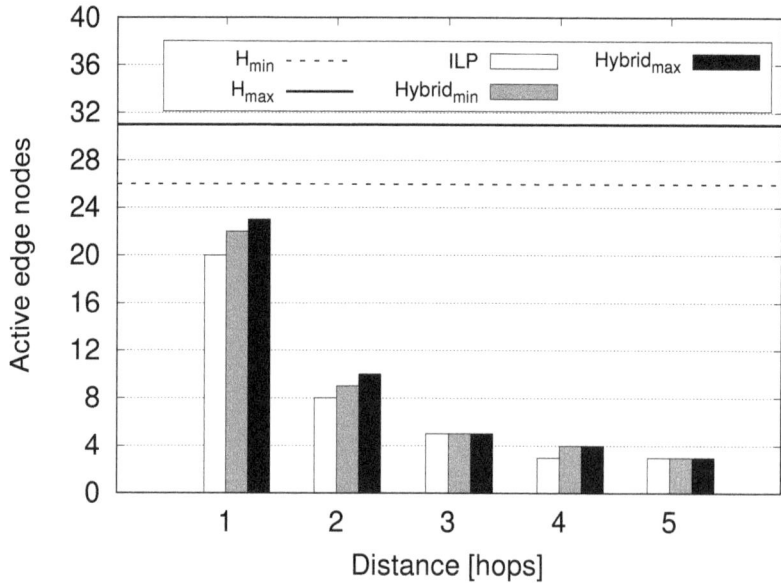

Figure 5. Total number of active edge nodes as a function of the allowed distance between RRUs and edge nodes for network N_{38}: Maximum and minimum costs of the hybrid results are reported after both phases.

In Figure 6, the gain of centralization of BBU and edge cloud functionalities is presented as a function of the allowed distance from RRUs by comparing the ILP results with the results of the hybrid approach at the end of phase 1 (denoted as H) and phase 2 in the maximum-cost case. This gain is relevant both for ILP and hybrid, with the hybrid being very close or coincident to the optimal solution. In the worst case (i.e., distance constraint equal to 1 hop), the hybrid provides only 8% gain reduction. As expected, phase 1 provides only suboptimal solutions. It is, therefore, evident the role of phase 2 of the hybrid approach in achieving a high centralization gain with respect to the plain coverage achieved in phase 1.

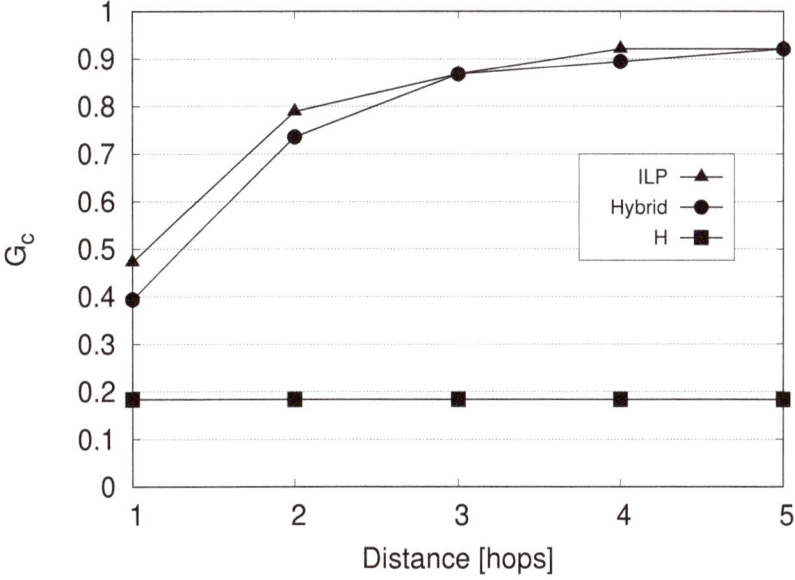

Figure 6. Centralization gain as a function of the allowed distance between RRUs and edge nodes for network N_{38}: Results are reported for the maximum cost for hybrid (phase 1 and phase 2), and ILP.

Table 2 reports the number of active links, wavelengths over the most used link, and overall wavelengths in network N_{38} for the two strategies. By comparing the strategies, it is possible to observe that the ILP requires a slightly higher number of wavelengths with respect to the hybrid approach when the number of active nodes is lower (distance constraints 1, 2, and 4). Nevertheless, because the activation cost of a node is much larger than the cost of a wavelength, the ILP solution always reaches a lower cost solution compared with the hybrid approach. When the ILP and hybrid require the same amount of active nodes (distance constraints 3 and 5) the ILP requires fewer wavelengths than the hybrid approach due to a wider set of choices. This happens for similar reasons also for the wavelengths required over the most used link.

To solve the harder instances of the problem, the ILP takes 2.8 s, 22.75 s, and 10,010.17 s in the network N_{14}, N_{20}, and N_{38}, respectively, showing an increased computational complexity when the size of the problem increases. Solving the ILP with the hybrid approach instead allows to reduce the solving times to 2.2 s, 17.99 s, and 3647.88 s in the three networks due to the reduction of the solution space. It should be noted that, in order to see the differences between the two strategies, the evaluations proposed here are done for networks suitable to cover a small- or medium-sized city. In larger scenarios (i.e., networks with more edge nodes and links), it is not always possible to ensure a solution with the ILP approach. These scenarios can be instead tackled with the hybrid approach, which has been shown to provide results close to optimality.

Table 2. Number of active links, wavelengths over the most used link, and total wavelengths for the hybrid and ILP for different distance constraints in network N_{38}.

Dist.	Hybrid			ILP		
[hops]	Active	Max	Total	Active	Max	Total
1	45	10	530	48	10	560
2	51	40	950	50	40	1040
3	49	70	1370	51	60	1350
4	52	70	1530	48	80	1830
5	51	80	1790	52	80	1780

7. Conclusions

This paper addresses the problem of providing low latency and reliable services in vehicular scenarios in a cost-efficient way using 4G and 5G networks. Baseband resources of C-RAN can be co-located with MEC resources to achieve target service requirements. An ILP model for the cost-efficient deployment of baseband and edge cloud resources with reliability against single node failure is proposed. In addition, a heuristic technique is also proposed to reduce computational complexity of the ILP model by proper selection of a subset of edge nodes for the optimization phase. Results show that the hybrid approach provides similar results to the ILP ones while considerably reducing the solving time.

Author Contributions: This paragraph specifies the individual contributions of authors to this article. Conceptualization, F.T., B.M.K., E.A., and C.R.; methodology and software, F.T. and B.M.K.; validation, F.T., B.M.K., and E.A.; investigation, E.A.; writing—original draft preparation, F.T., B.M.K., and E.A.; writing—review and editing, F.T., B.M.K., E.A., and C.R.

Funding: This research was carried out as a part of PhD activities and was partially funded by the Italian Ministry of Education, University and Research (MIUR).

Acknowledgments: The authors would like to thank Sebastiano Cucinotta for his contribution to the definition of the heuristic algorithm during the preparation of his Master project.

Conflicts of Interest: The authors declare no conflict of interest.

References

1. World Health Organization. *Global Status Report on Road Safety 2018*; Technical Report; WHO: Geneva, Switzerland, 2018.
2. GSMA. *Connecting Vehicles Today and in the 5G Era with C-V2X*; White Paper; GSMA: London, UK, 2019.
3. Mueck, M.; Karls, I. *Networking Vehicles to Everything: Evolving Automotive Solutions*; De Gruyter: Berlin, Germany, 2018.
4. 5G-TRANSFORMER Initial System Design–Project Grant No. 761536. Deliverable D1.2, H2020; 5G-TRANSFORMER. Available online: http://5g-transformer.eu/index.php/deliverables/ (accessed on 1 October 2018).
5. Checko, A.; Christiansen, H.L.; Yan, Y.; Scolari, L.; Kardaras, G.; Berger, M.S.; Dittmann, L. Cloud RAN for mobile networks A technology overview. *IEEE Commun. Surv. Tutor.* **2015**, *17*, 405–426. [CrossRef]
6. Van Lingen, F.; Yannuzzi, M.; Jain, A.; Irons-Mclean, R.; Lluch, O.; Carrera, D.; Perez, J.L.; Gutierrez, A.; Montero, D.; Marti, J.; et al. The Unavoidable Convergence of NFV, 5G, and Fog: A Model-Driven Approach to Bridge Cloud and Edge. *IEEE Commun. Mag.* **2017**, *55*, 28–35. [CrossRef]
7. Öhlén, P.; Skubic, B.; Rostami, A.; Fiorani, M.; Monti, P.; Ghebretensaé, Z.; Mårtensson, J.; Wang, K.; Wosinska, L. Data plane and control architectures for 5G transport networks. *J. Light. Technol.* **2016**, *34*, 1501–1508. [CrossRef]
8. Nobre, J.C.; de Souza, A.M.; Rosário, D.; Both, C.; Villas, L.A.; Cerqueira, E.; Braun, T.; Gerla, M. Vehicular software-defined networking and fog computing: Integration and design principles. *Ad Hoc Netw.* **2019**, *82*, 172–181. [CrossRef]

9. European Telecommunications Standards Institute. *Cloud RAN and MEC: A Perfect Pairing*; White Paper; ETSI: Sophia Antipolis, France, 2018.
10. Khorsandi, B.M.; Tonini, F.; Raffaelli, C. Design methodologies and algorithms for survivable C-RAN. In Proceedings of the 2018 International Conference on Optical Network Design and Modeling (ONDM), Dublin, Ireland, 14–17 May 2018; pp. 106–111. [CrossRef]
11. Liu, L.; Chen, C.; Pei, Q.; Maharjan, S.; Zhang, Y. Vehicular Edge Computing and Networking: A Survey. *arXiv* **2019**, arXiv:1908.06849.
12. 3GPP. *3rd Generation Partnership Project; Technical Specification Group Services and System Aspects; Summary of Rel-14 Work Items*; White Paper; 3GPP: Sophia Antipolis, France, 2018.
13. European Telecommunications Standards Institute. *Intelligent Transport Systems (ITS); Access Layer Specification for Intelligent Transport Systems Operating in The 5 GHz Frequency Band*; White Paper; ETSI: Sophia Antipolis, France, 2012.
14. Ning, Z.; Wang, X.; Huang, J. Mobile Edge Computing-Enabled 5G Vehicular Networks: Toward the Integration of Communication and Computing. *IEEE Veh. Technol. Mag.* **2019**, *14*, 54–61. [CrossRef]
15. Fiorani, M.; Skubic, B.; Mårtensson, J.; Valcarenghi, L.; Castoldi, P.; Wosinska, L.; Monti, P. On the design of 5G transport networks. *Photonic Netw. Commun.* **2015**, *30*, 403–415. [CrossRef]
16. Larsen, L.M.P.; Checko, A.; Christiansen, H.L. A Survey of the Functional Splits Proposed for 5G Mobile Crosshaul Networks. *IEEE Commun. Surv. Tutor.* **2019**, *21*, 146–172. [CrossRef]
17. Pfeiffer, T. Next generation mobile fronthaul and midhaul architectures. *J. Opt. Commun. Netw.* **2015**, *7*, B38–B45. [CrossRef]
18. 3GPP. *TR38.801—Radio Access Architecture and Interfaces*; Technical Report; 3GPP: Sophia Antipolis, France, 2017.
19. eCPRI V2.0 Specification. 2019. Available online: http://www.cpri.info/spec.html (accessed on 10 May 2019).
20. Shah, S.A.A.; Ahmed, E.; Imran, M.; Zeadally, S. 5G for Vehicular Communications. *IEEE Commun. Mag.* **2018**, *56*, 111–117. [CrossRef]
21. Zhao, Y.; Wang, W.; Li, Y.; Colman Meixner, C.; Tornatore, M.; Zhang, J. Edge Computing and Networking: A Survey on Infrastructures and Applications. *IEEE Access* **2019**, *7*, 101213–101230. [CrossRef]
22. Leyva-Pupo, I.; Santoyo-González, A.; Cervelló-Pastor, C. A Framework for the Joint Placement of Edge Service Infrastructure and User Plane Functions for 5G. *Sensors* **2019**, *19*, 3975. [CrossRef] [PubMed]
23. Santoyo-González, A.; Cervelló-Pastor, C. Latency-aware cost optimization of the service infrastructure placement in 5G networks. *J. Netw. Comput. Appl.* **2018**, *114*, 29–37. [CrossRef]
24. Musumeci, F.; Bellanzon, C.; Carapellese, N.; Tornatore, M.; Pattavina, A.; Gosselin, S. Optimal BBU placement for 5G C-RAN deployment over WDM aggregation networks. *J. Light. Technol.* **2016**, *34*, 1963–1970. [CrossRef]
25. Raffaelli, C.; Khorsandi, B.M.; Tonini, F. Distributed Location Algorithms for Flexible BBU Hotel Placement in C-RAN. In Proceedings of the 2018 20th International Conference on Transparent Optical Networks (ICTON), Bucharest, Romania, 1–5 July 2018; pp. 1–4. [CrossRef]
26. Khorsandi, B.M.; Tonini, F.; Raffaelli, C. Centralized vs. distributed algorithms for resilient 5G access networks. *Photonic Netw. Commun.* **2019**, *37*, 376–387. [CrossRef]
27. Shehata, M.; Musumeci, F.; Tornatore, M. Resilient BBU placement in 5G C-RAN over optical aggregation networks. *Photonic Netw. Commun.* **2019**, *37*, 388–398. [CrossRef]
28. Khan, A.A.; Abolhasan, M.; Ni, W. 5G next generation VANETs using SDN and fog computing framework. In Proceedings of the 2018 15th IEEE Annual Consumer Communications Networking Conference (CCNC), Las Vegas, NV, USA, 12–15 January 2018; pp. 1–6. [CrossRef]
29. Wang, S.; Zhao, Y.; Xu, J.; Yuan, J.; Hsu, C.H. Edge server placement in mobile edge computing. *J. Parallel Distrib. Comput.* **2019**, *127*, 160–168. [CrossRef]
30. Tinini, R.I.; Batista, D.M.; Figueiredo, G.B.; Tornatore, M.; Mukherjee, B. Low-latency and energy-efficient BBU placement and VPON formation in virtualized cloud-fog RAN. *IEEE/OSA J. Opt. Commun. Netw.* **2019**, *11*, B37–B48. [CrossRef]
31. Fiorani, M.; Rostami, A.; Wosinska, L.; Monti, P. Abstraction models for optical 5G transport networks. *IEEE/OSA J. Opt. Commun. Netw.* **2016**, *8*, 656–665. [CrossRef]

32. European Telecommunications Standards Institute. *MEC in 5G Networks*; White Paper; ETSI: Sophia Antipolis, France, 2018.
33. IBM. IBM ILOG CPLEX Optimization Studio V12.6.3. 2018. Available online: https://www.ibm.com/support/pages/downloading-ibm-ilog-cplex-optimization-studio-v1263 (accessed on 1 October 2018).

 © 2019 by the authors. Licensee MDPI, Basel, Switzerland. This article is an open access article distributed under the terms and conditions of the Creative Commons Attribution (CC BY) license (http://creativecommons.org/licenses/by/4.0/).

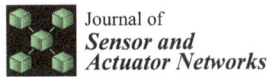

Review

The Use of Meta-Surfaces in Vehicular Networks

Barbara M. Masini [1,*], Cristiano M. Silva [2] and Ali Balador [3]

1. CNR-IEIIT, v.le Risorgimento, 2, 40136 Bologna, Italy
2. Departamento de Tecnologia, Universidade Federal de São João del-Rei, São João del-Rei 36307-352, Brazil; cristiano@ufsj.edu.br
3. Malardalen University, 1, 722 20 Vasteras, Sweden; ali.balador@ri.se
* Correspondence: barbara.masini@ieiit.cnr.it

Received: 30 November 2019; Accepted: 27 February 2020; Published: 2 March 2020

Abstract: Mobility as a service is becoming a new paradigm in the direction of travel planning on the basis of the best service offered by the travelled roads. Hence, the environment in which people move will become smarter and more and more connected to grant services along the whole path. This opens new challenges related not only to the on board connectivity and wireless access technologies, but also on the reliability and efficiency of the surrounding environment. In this context, reconfigurable meta-surfaces play a crucial role, since they can be used to coat buildings, vehicles or any other suitable surfaces and let the environment become an active part of the communication system by opportunistically redirecting (i.e., reflecting, without generating new waves) signals to the target receivers. The objective of this paper is to highlight the limits of current wireless access technologies for vehicular scenarios and to discuss the potential impact of a smart environment made of reconfigurable meta-surfaces on some next generation vehicular use cases, such as cooperative driving and vulnerable road users (VRUs) detection. In addition, a preliminary model is presented to derive, in a simplified way, the performance of an IEEE 802.11p network in terms of collision probability. Even if analytical and based on simplified assumptions, this model has been validated through simulations and allows to compare the performance of the network with and without reconfigurable meta-surfaces.

Keywords: connected vehicles; reconfigurable meta-surface; smart environment; cooperative driving; vulnerable road user detection; collision probability

---

1. Introduction

Listening to Alexander Lautz from Deutsche Telekom, the device that, 10 years from now, we will look back at as the device of the 5G era, will be the car [1]. This is not just a catchphrase, but it is a clear index of the new perception of mobility as a service and not only as a transportation from an origin to a destination. Today, our on board smart navigator suggests the best route as the shortest one (in time or distance); tomorrow the mobility manager (on board or distributed in the cloud) will plan our commutes on the best connected roads or following the combination of routes and transportation that better fulfill the user's preference and needs. The concept of Internet of vehicles (IoV) has recently emerged just to indicate the need to go beyond the potentialities of vehicular ad-hoc networks (VANETs) and pass from the concept of smartphone to that of smart car, that is a moving connected object in a connected environment [2–5].

To obtain reliable service along the travel route, we need not only connected and automated car, but also a smart environment, able to opportunistically contribute to the signals propagation and to an ubiquitous and reliable coverage.

Different wireless access technologies are running to come on the market of connected vehicles for vehicle-to-everything (V2X) communications [6,7]: on the one hand, the old fashioned IEEE 802.11p (or its European version ITS-G5) is only waiting for a mandatory and clear business model to be installed on board of all new vehicles [8,9] and, on the other hand, the newer cellular-V2X (C-V2X) proposed by 3GPP Release 14 is rushing into things promising better coverage, higher throughput and lower latency with respect to its competitor [10,11]. Recent works studied the performance of both IEEE 802.11p and C-V2X in different realistic scenarios, demonstrating their potentialities in terms of connectivity, packet reception ratio, latency and coverage, but also highlighting some limits in obstructed scenarios or congested roads [12–21].

What could be enhanced, looking forward, is the environment: a smart environment could drastically enhance the performance of wireless access technologies and having, as a consequence, an impact on connected vehicles related applications. Motivated by this, researchers from both the academical world and the industrial one, are proposing new solutions to smarten up the cities, starting from the street, buildings and citizen themselves [1]. This includes sensors embedded in the roadway [22], wireless access technologies on traffic lights or lamps along the roads [23], vehicular social networks [24], vehicles' routing [25–27], management of vehicular communication [28,29], cameras, smart wearable devices, etc. In this context, for example, the pilot project Austria's Autobahn uses Cisco's devices to connect tens thousands sensors with the objective to monitor traffic and road conditions. This represents an important example of how cities and public administrations move toward smart connected environments to improve safety, traffic efficiency, road capacity and infotainment [30,31]. The interest in smart environment is demonstrated also by the 6G Wireless Flagship Program (the world's first 6G research program), which indicates beside new wireless communication and computer science topics, also the importance of new electronics materials.

However, in spite of the huge effort in this direction, there will still scenarios in which the communication is obstructed by strong obstacles, preventing good links and allowing poor performance in terms of data rate, error rate, coverage and latency. In this context, reconfigurable meta-surfaces can play an active role, opportunistically redirecting the radio waves to improve connectivity and enabling the establishing of new and potentially stronger links [32]. Meta-surfaces are thin electromagnetic meta-material with typically sub-wavelength thickness and large in transverse size [33,34]. They are composed of sub-wavelength scattering particles that can revise the Snell's law redirecting the radio waves in the desired direction and can do this run time, changing the redirection of the waves time by time, according to the generalized Snell's laws, thus providing different values for the angles of incidence and reflection. Beyond meta-surface, what it is really challenging and stimulating, is the use of a *reconfigurable* meta-surface, where the scattering particles are not fixed, but can be moved and modified depending on the input they receive from the external world [35,36]. The idea of reconfiguring the wireless propagation environment has emerged only recently with focus especially on the indoor environments, where reconfigurable meta surfaces become connected to the rest of the scenario interacting with the connected objects and serving the user needs in unprecedented ways [37,38].

Several works deal with antenna design and performance optimization, such as [39], where a meta-superstrate for two vertically polarized MIMO antenna elements at the base-station is proposed to reduce the inter-element spacing. Preliminary evaluations related to the use of meta-surfaces in outdoor scenarios are related to the proposal of algorithms to minimize the total transmit power at the base station of a cellular system conditioned to the users quality of service (QoS) constraints [40] or to maximize either the energy or the spectral efficiency of a reconfigurable meta-surface multi-user MISO system [41]. Instead of using sensors embedded in the roadway and on traffic lights, reconfigurable meta-surfaces could be exploited to extend coverage in the highly dynamic vehicular environment, by coating the environment with intelligent meta materials.

Hence, the objective of this work can be summarized as follows:

- To describe the main characteristics of meta-surfaces.
- To highlight potential uses of reconfigurable meta-surfaces especially when adopted in vehicular environments. In this context, two main use cases are considered: cooperative driving and vulnerable road users (VRUs) detection.
- To demonstrate, through a simple analytical model (validated by simulation), the improvement that a reconfigurable meta-surface can provide in reducing the collision probability when random access to the medium is adopted for vehicle-to-vehicle (V2V) communications.

This work is organized as follows: in Section 2, the main advantages and limits of the two main candidates radio access technologies for vehicular networks are highlighted; in Section 3, the concept of reconfigurable meta surface is introduced and the use of reconfigurable meta-surfaces in vehicular networks is proposed and discussed especially referring to two case studies, cooperative driving and pedestrian detection. A simplified model for the evaluation of the impact of meta-surfaces on the IEEE 802.11p performance in terms of collision probability is presented in Section 4 and is validated by simulations. Finally, in Section 5 our conclusions are drawn.

2. Technologies for Vehicular Networks

Nowadays, two are the candidate enabling technologies for vehicular communications: IEEE 802.11p (or its European version ETSI ITS G5) and C-V2X.

IEEE 802.11p dates back in 2004 when the IEEE 802.11 working group started a discussion on how to modify and adapt Wi-Fi for dynamic environment, reducing the signaling and overhead of the nomadic version to support a completely different environments. It was then standardized in 2010 and tested in different field trials all around the world also with thousands of vehicles, demonstrating good performance in different use cases, thus representing a commercially available technology. On the other hand, C-V2X has been defined by 3GPP in 2016 within long term evolution (LTE) Release 14 and frozen in 2017 with the first plug test that took place in December 2019 in Malaga, Spain, demonstrating the 95% of success in terms of interoperability issues [42]. Hence, it can be observed that, while the IEEE 802.11p community spent several years from the first discussions to standardization and then on how and when set it up it on board, the cellular world sprinted forward and in a couple of years, test devices are ready for interoperability tests using a widespread technology, as it is demonstrated by the summary of timeline in Figure 1, with different colors showing the timeline of the different technologies. As it can be observed, the speed of development of C-V2X is much higher than Wi-Fi for mobility (IEEE 802.11p) and, despite being frozen in 2017, it will be ready for commercial installation on boards from 2020.

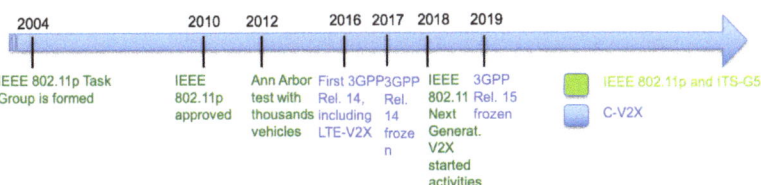

Figure 1. Summary of timeline of Wi-Fi for mobility and cellular vehicle-to-everything (C-V2X), showing the different sprint of the two standardization and experimentation processes.

2.1. IEEE 802.11p

IEEE 802.11p defines the physical (PHY) and medium access control (MAC) layer protocols. At the PHY layer, IEEE 802.11p adopts orthogonal frequency division multiplexing (OFDM) with 52 subcarriers of which 48 used for data and 4 for pilots. The OFDM symbol lasts 8 µs and the subcarrier spacing is 156.25 kHz, bringing to a raw bandwidth of 10 MHz. Eight modulation and coding schemes (MCSs) are possible, with modulations going from BPSK to 16-QAM with convolutional coding. At the MAC layer, carrier sensing multiple access with collision avoidance (CSMA/CA) is adopted, hence, when a node has to transmit a packet, it senses the medium; if the medium is idle the packet is transmitted after an Arbitration inter-frame spacing (AIFS) interval time (that takes into account potential delays in the propagation due to distant nodes), otherwise a mechanism based on random backoff is performed to reduce the probability of collisions by letting nodes to randomly start the next sensing phase for transmission. In addition, in the vehicular scenario, the acknowledgement and request to send/clear to send—RTS/CTS—mechanism are not foreseen to accelerate connection.

CSMA/CA has the advantage of being completely distributed and does not need any synchronization procedure, but, on the other hand, it suffers from collisions in dense vehicular environments [43], thus mechanisms to avoid overloads are necessary, such as decentralized congestion control (DCC) algorithm proposed by ETSI and SAE or new algorithm proposed in the literature [44], such as full duplex carrier sensing multiple access with collision detection (CSMA/CD) mechanism [45].

2.2. C-V2X

LTE Release 14 represents the first solution in the history of cellular world which includes direct resource allocation and communication without cellular coverage provided by an eNodeB with devices in high relative mobility. LTE-V2X is also called sidelink (since not downlink or uplink, but direct V2V) and its communication interface is named PC5. At the PHY layer, LTE-V2X adopts OFDM as IEEE 802.11p, whereas at the MAC layer it is based on the access used in the uplink of LTE, hence single carrier frequency division multiple access (SC-FDMA), that consists in a sort of pre-distorsion of the OFDM signal to behave as a single carrier in the presence of non linear effects. However, resources are orthogonally allocated in the frequency-time matrix as for orthogonal frequency division multiple access (OFDMA). Specifically, in the frequency domain, the subcarrier spacing is 15 kHz and subcarriers are allocated in groups of 12 (i.e., 180 kHz); in the time domain, 14 symbols form a subframe of 1 ms, also called transmission time interval (TTI), and include 9 data symbols, 4 demodulation reference signal (DMRS) symbols, and 1 empty symbol for Tx-Rx switch and timing adjustment. LTE-V2X has a high number of MCSs, with 4-QAM and 16-QAM modulations and an almost continuous coding rate.

Two different *modes* are defined by 3GPP for resources allocation, Mode 3 and Mode 4 [10]. In both modes communication is in direct (ad-hoc) mode between two vehicles, but they differ in the way resources are allocated. In Mode 3, also known as controlled allocation, the resources are defined and allocated by the network, hence, it can be adopted when vehicles are in the coverage of an eNodeB. In Mode 4, also known as autonomous or out-of-coverage allocation, each node selects the resources to use for communication based on a sensing procedure and an semi-persistent scheduling (SPS) mechanism [12,46], hence it is fully distributed as IEEE 802.11p.

2.3. Challenges and Limits of Current Technologies

Since Release 14 was defined, several studies have been carried out to compare IEEE 802.11p and C-V2X: many works push C-V2X as a more efficient technology, supported by an already deployed architecture diffused worldwide and a clearer road-map for future evolutions. The 5G Automotive Association (5GAA) pushes the 5th generation C-V2X to simplify the communication among the different road users, whereas the Car-to-Car Consortium (C2C-CO) claims that a largely tested and consolidated technology such as Wi-Fi for mobility could be the starting point for future vehicular communications. In [11], comparative experiments with real devices are shown, demonstrating longer ranges for C-V2X with respect to IEEE 802.11p and verifying that the latency in C-V2X under congested conditions does not exceed 100 ms. Other works, instead, push IEEE 802.11p since it represents a widely tested and reliable technology, already available on the market [47].

IEEE 802.11p suffers from collisions due to random access to the channel. Since each node transmits only if the medium is sensed as idle after a random backoff interval time, collisions typically arise in dense scenarios when an increasing number of vehicles try to access the medium and transmit. This mechanism may cause a high resources wasting since a transmission needs to come to an end before a collision is recognized, even if the collision happened at the very beginning. Hence, the number of terminals hidden to each others increases, since two nodes transmitting to the same destination but are far from each other may not sense the reciprocal interference. Another problem is constituted by the exposed terminal problem that happens when two different transmissions toward two different destinations senses each other and defer the transmission even if they could occupy the channel.

Hence, IEEE 802.11p provides satisfactory performance for most vehicular applications (especially those typically requiring a latency around 100 ms), when the vehicular density is moderate. When the vehicular density increases, IEEE 802.11p performance rapidly deteriorates due to packet collisions. Another limit to the diffusion of IEEE 802.11p is also the need of road side units (RSUs) development along the roads, but this is out of the scope of the present work [48].

The access is more robust in C-V2X networks, where orthogonal resources are allocated (by the eNodeB in Mode 3 or autonomously in Mode 4) to reduce interference to the minimum. Several works show that the performance of C-V2X sidelink Mode 4 is higher with respect to that of IEEE 802.11p in terms of signal to noise ratio (SNR) and coverage. However, when the traffic density increases, the performance of C-V2X, too, drops rapidly, particularly for Mode 4 resulting in an increased interference level among C-V2X users [12].

Given the fast development of C-V2X, also IEEE formed a new Task Group in January 2019, named IEEE 802.11bd to study the evolution of 802.11 for next-generation V2X communications. The attention was specifically focused on acOFDM numerology re-design, multiple input multiple output (MIMO) techniques, advanced channel coding, and better pilots placing, Dual carrier modulation (DCM) and 20 MHz channels (hence, band doubled with respect to the actual release).

Also, 3GPP is working for next generation systems and in September 2019 Release 16 defined new architectural modification to support advanced V2X applications with more stringent QoS requirements compared to applications that can be supported by C-V2X. Release 16 is quite specific in defining use-cases requirements and allows new solutions that support a latency lower than 3 ms with a reliability of 99.999% [22,49]. The design of two different and contemporary wireless access technologies, if from the one hand presents regional regulators and auto-manufacturers with two options, on the other hand, it results in challenging spectrum management issues.

In addition to the technological limits of each technology, the wireless propagation environment in vehicular networks suffers from highly dynamical scenario, of sudden obstructions, complex intersections, etc. [50]. Hence, independently on which technology will be set up first on board, the environment still play a severe propagation role for vehicular connectivity, providing absence of control over the wireless propagation environment and, often, as a consequence, high power consumption of the wireless interface. It is in this context that reconfigurable meta-surfaces could help in reflecting or refracting the impinging waves opportunistically to improve the link QoS and reach a higher number of users/vehicles with a target key performance indicator (KPI).

3. Reconfigurable Meta-Surfaces in Vehicular Scenarios

3.1. Reconfigurable Meta-Surfaces: Generalities

A meta-surface is an artificial surface made by the repetition of an electromagnetic element, the meta-atom, to provide properties that cannot be found in natural materials. The major difference between a surface and a meta-surface relies in the properties of the latter of shaping the radio waves according to the generalized Snell's laws of reflection and refraction, providing a spatial phase variation with sub-wavelength resolution to control the direction of wave propagation and the shape of wavefront [51].

When meta-atoms are static, we have a static interaction with electromagnetic waves, but if meta-atoms incorporate phase switching components (such as MEMS or CMOS transistors), the scattering particles are not fixed and we can have that impinging waves are routed according to provide customized reflections and the scattering particles can be modified depending on the stimuli that the meta-surface receives from the external world. Hence, very small reflecting elements can be used to coat natural surfaces and smartly reconfigure the signal propagation for performance enhancement.

In [33], two different types of reconfigurable meta-surfaces are discussed: the first type is represented by coating surfaces for walls, buildings, etc., that can be managed by a network operator as software-defined radio, with the main objective to improve the network coverage; the second type is the one than can be embedded into objects, as for wearable devices for health monitoring, that can backscatter the impinging radio waves to relay the acquired information. In both cases, it is worth noting that reconfigurable meta-surfaces allow wireless network operators to offer new services without emitting additional radio waves, by simply recycling those already existing for other purposes.

The specific arrangements of the scattering particles determine how the meta-surface transforms the incident wave into arbitrary specified reflected and transmitted radio waves. As shown, for example, in Figure 2, the two vehicles cannot communicate neither through V2V communication, due to the presence of a truck made in large part of iron-based materials, nor through vehicle-to-infrastructure (V2I) communication due to high and large buildings reducing the link QoS to unacceptable levels. However, the presence of an additional element, coated with a reconfigurable meta-surface, allows the redirection of the waves to let the link between the two vehicles possible, transforming the environment from adversary to collaborator, improving the coverage and making the propagation conditions favorable. We can also think to the particles of the meta-surface as embedded antennas constituted by a planar array of a large number of reconfigurable passive elements, where each element is able to independently introduce a certain phase shift onto the incident electromagnetic waves.

Figure 2. An opportune obstacle coated with meta material to redirect the electromagnetic waves toward the target receiver, which could not be reached through direct link due to an obstacle.

This can yield extended coverage range with higher SNR levels with respect to the natural environment, in some cases also doubling the coverage distance with the same SNR level of natural environment [52].

In addition, reconfigurable meta-surfaces acting as reflectors are not affected by self-interference and by noise amplification effects since they do not act as relays and are not affected by such impairments [33]. However, beside their incredible properties, meta-surfaces also present some drawbacks. For example, the smart environment is more sensitive to channel estimation errors with performance that deteriorates much more than the classical system as the channel training SNR decreases. Moreover, the channel estimation errors are more significant as the user moves away from the reconfigurable meta-surface.

Another critical issue is constituted by the amount of sensed data that the meta-surfaces need to gather and reflect to be able to configure and optimize the environment following the network requirements. It has also to be observed that meta-surfaces reflect the impinging waves and the reflected rays are phase shifted and delayed. The phase shift and the delays introduced depend in a large part on the intrinsic factor of the surfaces, such as absorption and dielectric properties. Since latency is a critical requirement to satisfy, especially in next generation safety applications, these are characteristics that have to be taken into consideration during the engineering process to satisfy the QoS of a given environment [51]. Last, but not least, how to integrate reconfigurable meta-surfaces in future wireless networks is still an open issue.

Table 1 summarizes the main advantages and drawbacks or challenges of reconfigurable meta-surfaces.

In spite of these challenges, reconfigurable meta-surfaces play a crucial role in interconnecting the physical and the digital worlds in a seamless and efficient manner [33].

Table 1. Main advantages and challenges or drawbacks of reconfigurable meta-surfaces.

Pros	Cons
Enlarge the concept of software networks	Only prototypes are currently available
No generation of new signals but reuse of existing ones	Not immediate integration in wireless networks
Programmable frequency selection	Performance of wireless networks with meta-surfaces still under investigation
Potential increasing of information reliability	More sensitive to channel estimation errors
Sensing capabilities	Potential need of power sources
Storage capabilities	Potential need of storage capabilities
Deployment scalability	Can introduce delays by storing and releasing the reflected signals
Can offer new services without emitting additional radio waves	Reflected waves are phase shifted and delayed

3.2. Reconfigurable Meta-Surfaces for Enhanced Vehicular Scenarios

The propagation environment in vehicular networks is constituted by a set of physical fixed and mobile objects that affect the propagation of electromagnetic waves between the communicating devices, often causing detrimental effects on the communication process. It is outside the network control. Software reconfigurable meta-surfaces (obtained, for example, by coating the walls of the buildings between the communicating devices) can mitigate the negative effect of the propagation environment by controlling the electromagnetic behavior of reflecting and refracting waves according to the generalized Snell's law, i.e., the angles of incidence and reflection of the radio waves can be different and depend on the phases induced by the elements of the reconfigurable meta-surface [52].

Looking, for example, at Figure 3, two vehicles approaching a highly obstructed intersection from two different directions (south and east in Figure 3a), could not be aware one of each other. But, if the buildings were coated with reconfigurable meta surfaces remote programmable as in Figure 3b, the electromagnetic waves could be opportunistically refracted toward the desired target receiver, improving safety applications, both in terms of link QoS and latency.

(a) Buildings with natural surfaces.　　(b) Buildings coated with reconfigurable meta-surfaces.

Figure 3. The impact of meta-surfaces in obstructed intersections.

In Table 2, the possible uses of reconfigurable meta-surfaces are listed and some enabled or improved applications in vehicular scenarios are summarized.

Table 2. Potential uses of reconfigurable meta-surfaces when applied to vehicular applications.

Potential Uses of Reconfigurable Meta-Surfaces	Safety	Non-Safety
Beamforming	Incident detection Hazardous warning Cooperative collision avoidance Trajectories alignment	Video Sharing
Range extension	Information sharing for automated driving Precise long horizon information Cooperative platooning Cooperative driving	Info-traffic sharing
Uplink bottleneck resolution for V2I	Remote driving information uploading	Traffic information uploading in dense environments
Positioning	V2R detection V2P communication	Personalized information Location aware information
Remote sensing	Hazard prevention	Environment detection Extended sensors

In the following, we consider two main case studies where reconfigurable meta surfaces could provide important performance improvements: cooperative driving and pedestrian detection.

3.3. Cooperative Driving

Cooperative driving is listed among the challenging future case studies that next wireless access technologies have to address. Basically, actual applications that could be already available through IEEE 802.11p or LTE-V2X communication systems, are represented by the so called Day 1 and Day 1.5 cooperative-ITS service list [53] and are mostly based on the exchange of periodic beacon messages among vehicles to enable the awareness of the environment (e.g., probe vehicle data, traffic information, smart routing, etc.). In the future, vehicles will not only rely on information exchange related to actual position, speed, acceleration, etc., but they will also need to share intentions: this will allow each vehicle to have a glimpse into the future of other vehicles, allowing a human driver or an artificial on board intelligence to take the best decision/behavior.

5GCAR has identified five relevant use case classes: cooperative maneuver, cooperative perception, cooperative safety, autonomous navigation, remote driving [1]. Focusing on cooperative maneuver, the principle is, from the one hand, the sharing of local awareness and driving intentions and, on the other hand, the negotiation of the planned trajectories. This way, the driving trajectories can be coordinated and even optimized in a centralized or decentralized manner. In summary, cooperative driving systems are based on algorithms that control the vehicle behavior based on the behavior of the surrounding vehicles and allow to extend the perception range beyond line-of-sight and sensing angles. Examples of cooperative maneuver are cooperative lane merging, cooperative intersection crossing, platooning, etc. [54].

Today, for example, the 75% of lane changing accidents occur because of a lack of perception of the surrounding environment [55]. IEEE 802.11p and C-V2X Release14 cannot fully address cooperative driving requirements in terms of latency and reliability C-V2X Release 15 will incorporate 5G new radio (NR) features, providing higher data rate, ultra-low latency and higher reliability so that advanced use cases, such as cooperative driving, can be adddressed. In addition, accurate positioning and ranging will be included to enable sub-meter positioning.

To enable cooperative driving, information coming from different sensors of different vehicles of different vendors should be properly merged in a fast and reliable manner and each information should be strictly related to its accurate position in time and space [56]. This also implies that, as more and more advanced sensors populate the cars and produce information to be exchanged for a reliable cooperation, the amount of information to be exchanged increases and the risk of errors on the transmission channel increases as well [57]. Hence, new approaches have to be conceived. An example of fully distributed cooperation system for lane merging is reported in Figure 4, highlighting the impact of an obstructing building. In case it is coated with reconfigurable meta-surface, the probability of correct packet reception between the two vehicles would result increased. This scenario implies that all vehicles are equipped with sensing capabilities, data processing units also including data merging and data fusion and artificial intelligent capabilities for decision making.

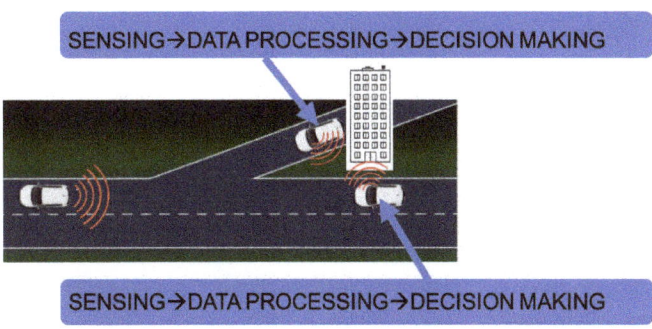

Figure 4. Example of cooperative driving scenario: lane merging.

3.4. Pedestrian Detection

In a document released by the European Commission in Brussels on 2019 April 4, it is possible to read that the European Union has some of the safest roads in the world. Nevertheless, more than 25,000 people still lose their lives on EU roads every year, and many more are seriously injured. In 2018, around 25,100 road fatalities were reported by the 28 EU Member States. This number is stilll very high, but it represents a first decrease of 21% compared to 2010.

A report provided by the National Highway Traffic Safety Administration (NHTSA) shows that over the past 40 years there has been a general downward trend in road mortal accidents, but focusing on the VRUs, in the year 2018, in the US 6283 pedestrians were killed, which is 3.4% higher that in 2017 and the highest number since 1990. Hence, in spite of safety measurements taken worldwide to reduce fatalities and injuries have lowered the number of accidents, these solutions do not protect the VRUs [58].

Hence, in addition to the safety measures to protect vehicle occupants, determined efforts have to be done to implement specific measures to prevent fatalities and injuries of VRUs, such as cyclists and pedestrians, is needed to protect users outside of the vehicle [59]. Current pedestrian detection is based on advanced driver assistance systems (ADAS) with onboard sensors, such as cameras, radar and LIDAR, to detect the presence of VRUs [60,61]. All of these sensors require line of sight (LOS) to properly work and they cannot detect VRUs in case of obstacles partially or totally occluding the visibility, such as trees, trucks or buildings.

The importance and actuality of this issue is also demonstrated by the fact that the U.S. Department of Transportation (USDOT) has released a summary of the available vehicle-to-pedestrian (V2P) technologies. Among these technologies, alert/notification to users (e.g., through mobile phone alert) is the only connected way considered. Several works have been published in the last years on this topic; some of these works, like [62,63], consider a IEEE 802.11p V2P communication, demonstrating a probability to inform the VRU around 0.8 with a beacon generation frequency at least equal to 1 Hz (for IEEE 802.11p, a unique Provider Service Identifier (PSID) to identify applications for the safety of VRUs is defined by standard [64]).

However, as stated in Section 2.3, IEEE 802.11p-based V2X communication suffers from congestions on the medium when the number of vehicles (and VRUs) increases. In urban centers, for examples, VRUs are present in large numbers leading to dense networks, whose performance quickly decreases. Adult VRUs have a smartphone in their pocket and, today, this is the easiest way to provide an alert. However, the actual embedded cellular technology (4G) does not allow V2X communication, yet, but only communications through the eNodeB, thus with a latency which could be critical for VRUs' safety applications [65]. When V2P communication is addressed, another important aspect to be carefully considered is positioning, since the global positioning system (GPS) accuracy and the accuracy of the combination of different sensors, even if suitable for different ranges and weather conditions, is not sufficient [66,67]. Actual cellular positioning also provides poor accuracy, but this could be (at least partially) overcome with the introduction of 5G wireless communication systems introducing MIMO systems at mmWave [68]. However, mmWave suffers from limited coverage ranges and are deeply attenuated by obstacles, thus they can help only in specific scenarios and typically short distances.

Therefore, new approaches are necessary trying to ensure the reliable delivery of safety messages. Figure 5 shows an example of how reconfigurable meta-surfaces can contribute to redirect messages to warn pedestrians hidden to an incoming vehicle. Note, in fact, that, in this case, radio communication can be drastically obstructed, but the coated building can however provide the timely delivery of the warning message.

Another issue to be addressed is related to the limited battery life of smartphones when used for V2P communications. In fact, in order to prevent accidents, both pedestrian smartphones and on board units (OBUs) need to frequently transmit beacons containing at least position, speed and direction. These periodic messages, even when sending small amount of data, can drain the smartphone battery very quickly [69,70]. This issue could be overcome, for example, with t-shirts coated with a reconfigurable meta-surface that, instead of transmitting a new signal consuming energy, simply reflects the received wave. It is also worth noting that some VRUs could not be connected at all, such as children. Hence, the protection and detection of VRUs represent another challenging and important scenario where a smart environment can act toward the improvement of safety. In fact, meta-surfaces can not only extend the coverage range, but also perform redirecting toward the right obstacle, alerting in case of real necessity.

4. The Impact of Meta-Surfaces: Performance Example

To provide numerical examples on the efficiency of meta-surfaces, we focus, in this Section, on IEEE 802.11p as enabling technology. To access the medium, IEEE 802.11p adopts CSMA/CA, hence, when a node needs to transmit a packet, it starts listening to the channel for an AIFS period, after which, if the channel is sensed idle, the packet is transmitted. If during AIFS the channel is sensed as busy or becomes busy, a random backoff algorithm is applied and the node will try to access the channel again after a random interval time chosen in the range [0–CW] [71]. Given this mechanism, IEEE 802.11p performance suffers from the hidden terminal effect and capture effect. As anticipated in Section 2.1, two nodes are hidden from each other when they are out of the reciprocal sensing range, but they are both transmitting toward the same destination. The capture effect, instead, is due to the presence of multiple transmitters that are sending their messages at the same time towards the same destination; in this case, it may happen that the power level of one of them is sufficiently higher than the interference received from the others to allow correct decoding at the receiver. Otherwise, the message is not captured.

To make some example evaluations, we assume a highway scenario with variable traffic conditions and multiple lanes, as shown in Figure 6. To analytically estimate the collision probability, this 2-D scenario is approximated with an 1-D scenario with the assumption of Poisson distributed vehicles with variable density ρ. Each vehicle is assumed to be equipped with an OBU that periodically transmits beacons of B_b bytes with a frequency f_b expressed in Hz.

Following the model presented in [15], it is possible to evaluate the probability to collide, either in visibility or in case of hidden terminals and not captured, as

$$p_c = 1 - (1 - p_{c\text{-vis}}) \cdot \left(1 - p_{c\text{-ht}} \cdot (1 - \frac{p_{\text{capt}}}{2})\right) \quad (1)$$

where $p_{c\text{-vis}}$ is probability that the considered vehicle (i.e., an OBU) senses the channel busy and ends with colliding with at least one other vehicle, p_{capt} is the capture effect probability (the capture effect avoids the collision if and only if the interference received from the other transmitting node is below a given threshold) and $p_{c\text{-ht}}$ is the probability to collide due to the presence of an hidden terminal either ending or starting its transmission during the observed communication (more details to derive the different probabilities can be found in [15]).

Figure 5. Pedestrian detection.

In general (i.e., without meta-surfaces), the probability that a collision occurs due to the presence of hidden terminals can be found as [15]:

$$p_{\text{c-ht}} = 1 - p_{\text{nc-h1}} \cdot p_{\text{nc-h2}} \qquad (2)$$

where $p_{\text{nc-h1}}$ is the probability that no hidden terminal is transmitting and $p_{\text{nc-h2}}$ is the probability that no hidden node starts its transmission during the observed communication.

Let us now assume that $p_{\text{c-vis}}$ and p_{capt} remain equal also in the presence of meta-surfaces, whereas $p_{\text{c-ht}}$ is affected by the presence of meta-surfaces which could limit the hidden terminal effect through the redirection of reflected paths that could extend the coverage, as shown in Figure 6. Hence, we assume the presence of at least one meta-surface coated object which allows to grant that no hidden node starts its transmission during the observed communication, always providing $p_{\text{nc-h2}} = 1$, simplifing (2) that reduces to

$$p_{\text{c-ht}} = 1 - p_{\text{nc-h1}}. \qquad (3)$$

Hence, the collision probability (1) becomes

$$p_c = 1 - (1 - p_{\text{c-vis}}) \cdot \left[1 - (1 - p_{\text{nc-h1}}) \cdot \left(1 - \frac{p_{\text{capt}}}{2}\right)\right]. \qquad (4)$$

This is possible through the extension of the coverage range of the transmitting node that allows the hidden terminal to hear for an ongoing transmission, as shown in Figure 6b. This is a simplified model, but help in individuating and evaluating the impact of the proposed solution on the wireless access technology performance.

In order to provide some numerical results, we assume each vehicle equipped with an OBU that periodically transmits beacons of $B_b = 300$ bytes every $f_b = 10$ Hz. Among the eight available modes of IEEE 802.11p, we consider mode 3, providing a raw data rate of 6 Mb/s.

To demonstrate the validity of the model simulation results are also provided, referring to a 16 km highway with 3 lanes per direction. The main settings are reported in Table 3. As far as the propagation is concerned, we assume the following path loss model

$$L(x) = L_0 \cdot x^\beta \qquad (5)$$

where L_0 is the path loss at the reference distance of 1 m, x is the distance between the transmitter and the receiver, and β is the path loss exponent. Hence, given a distance d between the transmitting and receiving vehicles and n_{int} interferers each at distance $d_{\text{int}}^{(i)}$ (with $i \in [1, n_{\text{int}}]$) from the receiver, all of them transmitting with the same power P_{tx}, the signal to noise and interference ratio (SINR) at the receiver can be calculated calculated as

$$\gamma = \frac{\frac{P_{\text{tx}} \cdot G_r}{L_0 \cdot d^\beta}}{P_n + \sum_{i \in [1, n_{\text{int}}]} \frac{P_{\text{tx}} \cdot G_r}{L_0 \cdot d_{\text{int}}^{(i)\beta}}} \qquad (6)$$

where G_r is the antenna gain at the receiver and P_n is the noise power. By inverting (6), the transmission range r_{tx} (i.e., without interference) can be obtained as

$$r_{\text{tx}} = \left[\frac{P_{\text{tx}} \cdot G_r}{L_0 \cdot P_n \cdot \gamma}\right]^{1/\beta}. \qquad (7)$$

Vehicles in the transmission range r_{tx} that are not visible to the transmitter are considered as hidden terminals.

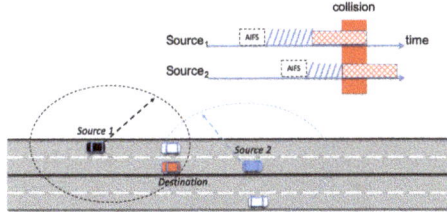 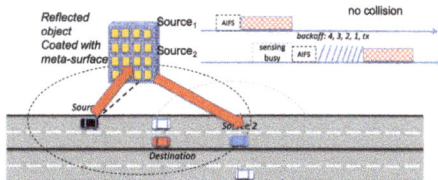

(a) Hidden terminal problem.

(b) The coverage range of Source$_1$ is extended by a meta-surface and Source$_2$ senses the channel occupied and defers its transmission.

Figure 6. The coverage range of Source$_1$ can be extended by a meta-surface to solve the hidden terminal problem.

Table 3. Main settings.

Parameter (Symbol)	Value
IEEE 802.11p mode	3
Modulation	QPSK
Coding rate	1/2
Raw data rate	6 Mb/s
Vehicle density (ρ)	Variable
Beacon frequency (f_b)	10 Hz
Vehicle density (ρ)	Variable
Equivalent radiated power (P_{tx})	33 dBm
Receiver antenna gain (G_r)	3 dB
Path loss at 1 m at 5.9 GHz (L_0)	47.86 dB
Path loss exponent (β)	2.75
Noise power over 10 MHz (P_n)	−95 dBm
Beacon size (B_b)	300 bytes
Contention window ($\hat{CW}$)	15
Slot duration (t_σ)	13 μs
AIFS duration (t_{aifs})	58 μs
Overhead per packet	40 μs

In Figure 7, the collision probability given by (4) is plotted as a function of the distance for different values of vehicular density ρ. A comparison between the cases with and without meta-surfaces is reported. In addition, simulation results are also provided showing a good agreement with the analysis. As expected, independently on the vehicles density and on the use of meta-surfaces, the collision probability increases with the distance: for short distances the impact of hidden terminals is negligible and the collision probability is almost constant, then the impact of hidden terminals and low visibility is higher and the collision probability increases.

As a consequence, the impact of the use of meta-surfaces on the collision probability is much more evident as the distance and the vehicles density increase. For example, when the vehicles density $\rho = 0.05$, the adoption of meta-surfaces can reduce the collision probability from 0.12 to around 0.06 at a distance of 350 m, whereas when $\rho = 0.2$, the collision probability has a higher impact, reducing the collision probability of around 42% from 0.41 to around 0.27, which is a more tolerable value.

Note that since meta-surfaces do not act as relays, they do not introduce new signals in the network, hence the level of congestion is not increased. They reflect signals already present in the network, extending their coverage. In this sense, interference could also be extended, but algorithms to cancel it or reduce it could be applied as in other software-defined networks.

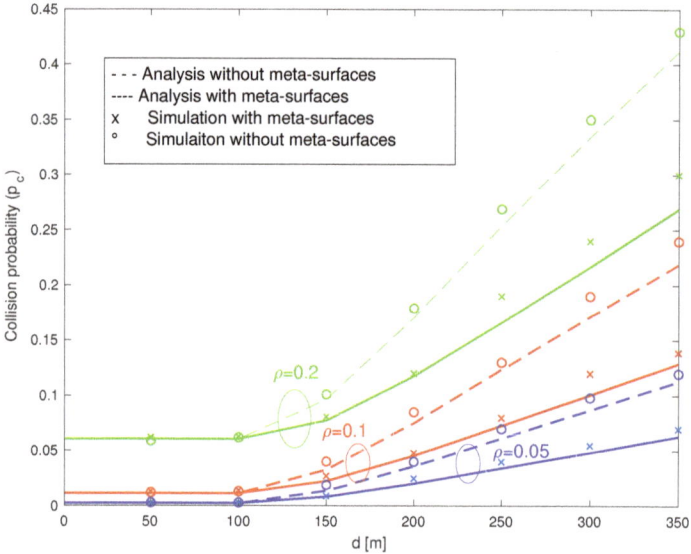

Figure 7. Collision probability as a function of the inter-vehicular distance with and without meta-surfaces varying the vehicles density ρ. Lines correspond to analysis, symbols to simulations.

5. Consclusions

In this work, we provided a general overview on how reconfigurable meta-surfaces can be used in vehicular scenarios by partially covering the limits of current wireless access technologies, such as IEEE 802.11p and C-V2X. Specifically, we aimed at underlining the main characteristics and potentialities of meta-surfaces when used in vehicular scenario, also providing a simple evaluation of their impact on network performance and demonstrating, in a simple scenario with Poisson random distributed vehicles, that if meta-surfaces can coat at least part of the scenario, they can extend the coverage range, limiting, for example, the hidden terminal problem that affects random access technologies. This model has been also validated by simulations on highways. Since meta-surfaces do not act as relays and do not introduce new signals in the network, the network congestion level is not affected and neither self interference nor amplification noise are added to the system.

Author Contributions: Conceptualization, methodology, software and writing original draft preparation: B.M.M.; literature review, writing review and editing, A.B. and C.M.S. All authors have read and agreed to the published version of the manuscript.

Funding: This research was partially funded by CNPq (Conselho Nacional de Desenvolvimento Científico e Tecnológico) grant 303933/2017-8, CAPES (Coordenação de Aperfeiçoamento de Pessoal de Nível Superior), and FAPEMIG (Fundação de Amparo à Pesquisa do Estado de Minas Gerais) grant APQ-02145-18.

Acknowledgments: The authors would like to thank Alessandro Bazzi and Alberto Zanella for helpful discussions and support, Marco di Renzo for his highlighting introduction to meta surfaces and Ian Akyildiz for his inspiring keynotes and never ending motivation.

Conflicts of Interest: The authors declare no conflict of interest.

References

1. On the Road to Self-Driving Cars, 5G Will Make Us Better Drivers. Available online: https://spectrum.ieee.org/telecom/wireless/mwc-barcelona-2019-on-the-road-to-selfdriving-cars-5g-will-make-us-better-drivers (accessed on 1 March 2020).

2. Yang, F.; Wang, S.; Li, J.; Liu, Z.; Sun, Q. An overview of Internet of Vehicles. *China Commun.* **2014**, *11*, 1–15. [CrossRef]
3. Kaiwartya, O.; Abdullah, A.H.; Cao, Y.; Altameem, A.; Prasad, M.; Lin, C.; Liu, X. Internet of Vehicles: Motivation, Layered Architecture, Network Model, Challenges, and Future Aspects. *IEEE Access* **2016**, *4*, 5356–5373. [CrossRef]
4. Aliyu, A.; Kaiwartya, O.; Cao, Y.; Lloret, J.; Aslam, N.; Usman, M. Towards Video Streaming in IoT Environments: Vehicular Communication Perspective. *Comput. Commun.* **2017**. [CrossRef]
5. Kumar, S.; Singh, K.; Kumar, S.; Kaiwartya, O.; Cao, Y.; Zhou, H. Delimitated Anti Jammer Scheme for Internet of Vehicle: Machine Learning Based Security Approach. *IEEE Access* **2019**, *7*, 113311–113323. [CrossRef]
6. Silva, C.M.; Masini, B.M.; Ferrari, G.; Thibault, I. A Survey on Infrastructure-Based Vehicular Networks. *Mob. Inf. Syst.* **2017**, *2017*, 28–56. [CrossRef]
7. Silva, C.M.; Silva, L.D.; Santos, L.A.L.; Sarubbi, J.F.M.; Pitsillides, A. Broadening Understanding on Managing the Communication Infrastructure in Vehicular Networks: Customizing the Coverage Using the Delta Network. *Future Internet* **2018**, *11*, 1. [CrossRef]
8. EUROPEAN COMMISSION. COMMISSION DELEGATED REGULATION (EU) of 13.3.2019 Supplementing Directive 2010/40/EU of the European Parliament and of the Council with Regard to the Deployment and Operational Use of Cooperative Intelligent Transport Systems. 2019. Available online: https://spectrum.ieee.org/telecom/wireless/mwc-barcelona-2019-on-the-road-to-selfdriving-cars-5g-will-make-us-better-drivers (accessed on 1 March 2020).
9. Autotalks and NXP Semiconductors. IEEE 802.11p Ahead of LTE-V2V for Safety Applications. 2017. Available online: https://www.nxp.com/docs/en/white-paper/LTE-V2V-WP.pdf (accessed on 1 March 2020).
10. Bazzi, A.; Cecchini, G.; Menarini, M.; Masini, B.M.; Zanella, A. Survey and Perspectives of Vehicular Wi-Fi versus Sidelink Cellular-V2X in the 5G Era. *Future Internet* **2019**, *11*, 122. [CrossRef]
11. 5GAA. V2X Functional and Performance Test Report; Test Procedures and Results. 2018. Available online: https://5gaa.org/wp-content/uploads/2018/10/FCC-USDOT-CV2X-v2.14_wo_Video-c1_Final.pdf (accessed on 1 March 2020).
12. Molina-Masegosa, R.; Gozalvez, J. LTE-V for Sidelink 5G V2X Vehicular Communications: A New 5G Technology for Short-Range Vehicle-to-Everything Communications. *IEEE Veh. Technol. Mag.* **2017**, *12*, 30–39. [CrossRef]
13. Min, W.; Winbjork, M.; Zhang, Z.; Blasco, R.; Do, H.; Sorrentino, S.; Belleschi, M.; Zang, Y. Comparison of LTE and DSRC-Based Connectivity for Intelligent Transportation Systems. In Proceedings of the IEEE 85th Vehicular Technology Conference (VTC Spring), Sydney, Australia, 4–7 June 2017; doi:10.1109/VTCSpring.2017.8108284. [CrossRef]
14. Nguyen, T.V.; Shailesh, P.; Sudhir, B.; Kapil, G.; Jiang, L.; Wu, Z.; Malladi, D.; Li, J. A Comparison of cellular vehicle-to-everything and dedicated short range communication. In Proceedings of the IEEE Vehicular Networking Conference (VNC), Torino, Italy, 27–29 November 2017.
15. Bazzi, A.; Masini, B.M.; Zanella, A.; Thibault, I. On the Performance of IEEE 802.11p and LTE-V2V for the Cooperative Awareness of Connected Vehicles. *IEEE Trans. Veh. Technol.* **2017**, *66*, 10419–10432. [CrossRef]
16. Cecchini, G.; Bazzi, A.; Masini, B.M.; Zanella, A. Performance comparison between IEEE 802.11p and LTE-V2V in-coverage and out-of-coverage for cooperative awareness. In Proceedings of the IEEE Vehicular Networking Conference (VNC), Torino, Italy, 27–29 November 2017; pp. 109–114. [CrossRef]
17. Vukadinovic, V.; Bakowski, K.; Marsch, P.; Garcia, I.D.; Xu, H.; Sybis, M.; Sroka, P.; Wesolowski, K.; Lister, D.; Thibault, I. 3GPP C-V2X and IEEE 802.11p for Vehicle-to-Vehicle communications in highway platooning scenarios. *Ad Hoc Netw.* **2018**, *74*, 17–29. [CrossRef]
18. Thota, J.; Abdullah, N.F.; Doufexi, A.; Armour, S. Performance of Car to Car Safety Broadcast Using Cellular V2V and IEEE 802.11P. In Proceedings of the IEEE Vehicular Technology Conference (VTC Spring), Porto, Portugal, 3–6 June 2018; pp. 1–5. [CrossRef]
19. Anwar, W.; Kulkarni, K.; Augustin, T.R.; Franchi, N.; Fettweis, G. PHY Abstraction Techniques for IEEE 802.11p and LTE-V2V: Applications and Analysis. In Proceedings of the 2018 IEEE Globecom Workshops (GC Wkshps), Abu Dhabi, UAE, 9–13 December 2018; pp. 1–7. [CrossRef]

20. Kühlmorgen, S.; Schmager, P.; Festag, A.; Fettweis, G. Simulation-based Evaluation of ETSI ITS-G5 and Cellular-VCS in a Real-World Road Traffic Scenario. In Proceedings of the IEEE Vehicular Technology Conference (VTC Fall), Chicago, IL USA, 27–30 August 2018.
21. Bastos, A.V.; Silva, C.M.; Silva, D., Jr. Assisted Routing Algorithm for D2D Communication in 5G Wireless Networks. In Proceedings of the 2018 Wireless Days (WD) (WD'18), Dubai, UAE, 3–5 April 2018; doi:10.1109/WD.2018.8361688. [CrossRef]
22. Silva, C.M.; Pitangui, C.G.; Miguel, E.C.; Santos, L.A.; Torres, K.B. Gamma-Reload Deployment: Planning the communication infrastructure for serving streaming for connected vehicles. *Veh. Commun.* **2020**, *21*, 100197. [CrossRef]
23. Silva, C.M.; Aquino, A.L.L.; Meira, W., Jr. Smart Traffic Light for Low Traffic Conditions. *Mob. Netw. Appl.* **2015**, 1–9. [CrossRef]
24. Oliveira, T.R.; Silva, C.M.; Macedo, D.F.; Nogueira, J.M.S. SNVC: Social networks for vehicular certification. *Comput. Netw.* **2016**, *111*, 129–140, doi:10.1016/j.comnet.2016.08.030. [CrossRef]
25. Silva, C.M.; Sarubbi, J.F.; Silva, D.F.; Porto, M.F.; Nunes, N.T. A Mixed Load Solution for the Rural School Bus Routing Problem. In Proceedings of the 2015 IEEE 18th International Conference on Intelligent Transportation Systems (ITSC), Las Palmas, Spain, 15–18 September 2015; pp. 1940–1945. [CrossRef]
26. Silva, C.M.; Guidoni, D.; Souza, F.S.; Pitangui, C.; Pitsillides, A. Using the Inter-Contact Time for Planning the Distribution of Roadside Units in Vehicular Networks. In Proceedings of the 19th IEEE International Conference on Intelligent Transportation Systems (ITSC 2016), Rio de Janeiro, Brazil, 1–4 November 2016.
27. Sarubbi, J.F.M.; Mesquita, C.M.R.; Wanner, E.F.; Santos, V.F.; Silva, C.M. A strategy for clustering students minimizing the number of bus stops for solving the school bus routing problem. In Proceedings of the NOMS 2016—2016 IEEE/IFIP Network Operations and Management Symposium, Istanbul, Turkey, 25–29 April 2016; pp. 1175–1180. [CrossRef]
28. Silva, C.M.; Meira, W., Jr. Design of roadside communication infrastructure with QoS guarantees. In Proceedings of the 2015 IEEE Symposium on Computers and Communication (ISCC), Larnaca, Cyprus, 6–9 July 2015; pp. 439–444. [CrossRef]
29. Silva, C.M.; Guidoni, D.L.; Souza, F.S.H.; Pitangui, C.G.; Sarubbi, J.F.M.; Pitsillides, A. Gamma Deployment: Designing the Communication Infrastructure in Vehicular Networks Assuring Guarantees on the V2I Inter-Contact Time. In Proceedings of the 2016 IEEE 13th International Conference on Mobile Ad Hoc and Sensor Systems (MASS), Brasilia, Brazil, 10–13 October 2016; pp. 263–271. [CrossRef]
30. Laplante, P. "Smarter" Roads and Highways. *IEEE Internet Things Mag.* **2018**, *1*, 30–35. [CrossRef]
31. Vegni, A.M.; Loscrí, V. A Survey on Vehicular Social Networks. *IEEE Commun. Surv. Tutor.* **2015**, *17*, 2397–2419. [CrossRef]
32. Basar, E.; Di Renzo, M.; De Rosny, J.; Debbah, M.; Alouini, M.; Zhang, R. Wireless Communications Through Reconfigurable Intelligent Surfaces. *IEEE Access* **2019**, *7*, 116753–116773. [CrossRef]
33. Renzo, M.D.; Debbah, M.; Huy, D.T.P.; Zappone, A.; Alouini, M.; Yuen, C.; Sciancalepore, V.; Alexandropoulos, G.C.; Hoydis, J.; Gacanin, H.; et al. Smart Radio Environments Empowered by AI Reconfigurable Meta-Surfaces: An Idea Whose Time Has Come. *arXiv* **2019**, arXiv:1903.08925.
34. Bukhari, S.; Vardaxoglou, J.; Whittow, W. A Metasurfaces Review: Definitions and Applications. *Appl. Sci.* **2019**, *9*, 2727. [CrossRef]
35. Liaskos, C.; Nie, S.; Tsioliaridou, A.; Pitsillides, A.; Ioannidis, S.; Akyildiz, I. A New Wireless Communication Paradigm through Software-Controlled Metasurfaces. *IEEE Commun. Mag.* **2018**, *56*, 162–169. [CrossRef]
36. Liaskos, C.; Nie, S.; Tsioliaridou, A.; Pitsillides, A.; Ioannidis, S.; Akyildiz, I.F. Realizing Wireless Communication through Software-defined HyperSurface Environments. *arXiv* **2018**, arXiv:1903.08925.
37. Liaskos, C.; Tsioliaridou, A.; Pitsillides, A.; Ioannidis, S.; Akyildiz, I.F. Using any Surface to Realize a New Paradigm for Wireless Communications. *arXiv* **2018**, arXiv:1903.08925.
38. Tan, X.; Sun, Z.; Jornet, J.M.; Pados, D. Increasing indoor spectrum sharing capacity using smart reflect-array. In Proceedings of the 2016 IEEE International Conference on Communications (ICC), Kuala Lumpur, Malaysia, 22–27 May 2016; pp. 1–6. [CrossRef]
39. Liu, F.; Guo, J.; Zhao, L.; Shen, X.; Yin, Y. A Meta-Surface Decoupling Method for Two Linear Polarized Antenna Array in Sub-6 GHz Base Station Applications. *IEEE Access* **2019**, *7*, 2759–2768. [CrossRef]
40. Wu, Q.; Zhang, R. Intelligent Reflecting Surface Enhanced Wireless Network: Joint Active and Passive Beamforming Design. *arXiv* **2018**, arXiv:1903.08925.

41. Huang, C.; Zappone, A.; Alexandropoulos, G.; Debbah, M.; Yuen, C. Large Intelligent Surfaces for Energy Efficiency in Wireless Communication. 2018. Available online: https://deepai.org/publication/large-intelligent-surfaces-for-energy-efficiency-in-wireless-communication (accessed on 1 March 2020).
42. C-V2X Plugtest. Available online: https://www.etsi.org/events/1659-cv2x-plugtests#pane-1/ (accessed on 1 March 2020).
43. Fallah, Y.P.; Huang, C.; Sengupta, R.; Krishnan, H. Analysis of Information Dissemination in Vehicular Ad-Hoc Networks With Application to Cooperative Vehicle Safety Systems. *IEEE Trans. Veh. Technol.* **2011**, *60*, 233–247. [CrossRef]
44. Balador, A.; Bohm, A.; Calafate, C.T.; Cano, J. A reliable token-based MAC protocol for V2V communication in urban VANET. In Proceedings of the 2016 IEEE 27th Annual International Symposium on Personal, Indoor, and Mobile Radio Communications (PIMRC), Valencia, Spain, 4–7 September 2016; pp. 1–6. [CrossRef]
45. Bazzi, A.; Campolo, C.; Masini, B.M.; Molinaro, A.; Zanella, A.; Berthet, A.O. Enhancing Cooperative Driving in IEEE 802.11 Vehicular Networks Through Full-Duplex Radios. *IEEE Trans. Wirel. Commun.* **2018**, *17*, 2402–2416. [CrossRef]
46. Bazzi, A.; Cecchini, G.; Zanella, A.; Masini, B.M. Study of the Impact of PHY and MAC Parameters in 3GPP C-V2V Mode 4. *IEEE Access* **2018**. [CrossRef]
47. NXP USA, I. A of the Comments of NXP USA, Inc., in the Letter before the Federal Communications Commission, Washington, DC. 2019. Available online: https://autoalliance.org/wp-content/uploads/2017/01/5.9-GHz-Comments.pdf (accessed on 1 March 2020).
48. Sarubbi, J.F.M.; Silva, T.R.; Martins, F.V.C.; Wanner, E.F.; Silva, C.M. Allocating Roadside Units in VANETs Using a Variable Neighborhood Search Strategy. In Proceedings of the 2017 IEEE 85th Vehicular Technology Conference (VTC Spring), Sydney, Australia, 4–7 June 2017; pp. 1–5. [CrossRef]
49. Naik, G.; Choudhury, B.; Park, J. IEEE 802.11bd & 5G NR V2X: Evolution of Radio Access Technologies for V2X Communications. *arXiv* **2019**, arXiv:1903.08925.
50. Loscrí, V.; Natalizio, E.; Costanzo, C. Simulations of the Impact of Controlled Mobility for Routing Protocols. *EURASIP J. Wirel. Commun. Netw.* **2009**, *2010*, 315381. [CrossRef]
51. Chen, H.T.; Taylor, A.; Yu, N. A review of metasurfaces: Physics and applications. *Rep. Prog. Phys.* **2016**, *79*. [CrossRef]
52. Nadeem, Q.U.A.; Kammoun, A.; Chaaban, A.; Debbah, R.; Alouini, M.S. Intelligent Reflecting Surface Assisted Multi-User MISO Communication. *arXiv* **2019**, arXiv:1906.02360.
53. A European strategy on Cooperative Intelligent Transport Systems, a Milestone towards Cooperative, Connected and Automated Mobility. 2016. Available online: https://ec.europa.eu/transport/themes/c-its_en (accessed on 1 Mar. 2020).
54. Hasan, S.; Balador, A.; Girs, S.; Uhlemann, E. Towards Emergency Braking as a Fail-Safe State in Platooning: A Simulative Approach. In Proceedings of the 2019 IEEE 90th Vehicular Technology Conference (VTC2019-Fall), Honolulu, HI, USA, 22–25 September 2019; pp. 1–5. [CrossRef]
55. Ammoun, S.; Nashashibi, F.; Laurgeau, C. An analysis of the lane changing manoeuvre on roads: The contribution of inter-vehicle cooperation via communication. In Proceedings of the 2007 IEEE Intelligent Vehicles Symposium, Istanbul, Turkey, 13–15 June 2007; pp. 1095–1100. [CrossRef]
56. Kim, S.; Qin, B.; Chong, Z.J.; Shen, X.; Liu, W.; Ang, M.H.; Frazzoli, E.; Rus, D. Multivehicle Cooperative Driving Using Cooperative Perception: Design and Experimental Validation. *IEEE Trans. Intell. Transp. Syst.* **2015**, *16*, 663–680. [CrossRef]
57. Wang, N.; Wang, X.; Palacharla, P.; Ikeuchi, T. Cooperative autonomous driving for traffic congestion avoidance through vehicle-to-vehicle communications. In Proceedings of the 2017 IEEE Vehicular Networking Conference (VNC), Amsterdam, The Netherlands, 14–16 November 2011; pp. 327–330. [CrossRef]
58. U.S. Department of Transportation (DOT), NHTSA Research Note. *2018 Fatal Motor Vehicle Crashes Overview*; U.S. Department of Transportation Report; NHTSA's National Center for Statistics and Analysis 1200 New Jersey Avenue SE: Washington, DC, USA, 2019.
59. Hussein, A.; Garcia, F.; Armingol, J.M.; Olaverri-Monreal, C. P2V and V2P communication for Pedestrian warning on the basis of Autonomous Vehicles. In Proceedings of the 2016 IEEE 19th International Conference on Intelligent Transportation Systems (ITSC), Rio de Janeiro, Brazil, 1–4 November 2016; pp. 2034–2039. [CrossRef]

60. Wu, X.; Miucic, R.; Yang, S.; Al-Stouhi, S.; Misener, J.; Bai, S.; Chan, W. Cars Talk to Phones: A DSRC Based Vehicle-Pedestrian Safety System. In Proceedings of the 2014 IEEE 80th Vehicular Technology Conference (VTC2014-Fall), Vancouver, WC, Canada, 14–17 September 2014; pp. 1–7. [CrossRef]
61. Tahmasbi-Sarvestani, A.; Nourkhiz Mahjoub, H.; Fallah, Y.P.; Moradi-Pari, E.; Abuchaar, O. Implementation and Evaluation of a Cooperative Vehicle-to-Pedestrian Safety Application. *IEEE Intell. Transp. Syst. Mag.* **2017**, *9*, 62–75. [CrossRef]
62. Anaya, J.J.; Merdrignac, P.; Shagdar, O.; Nashashibi, F.; Naranjo, J.E. Vehicle to pedestrian communications for protection of vulnerable road users. In Proceedings of the 2014 IEEE Intelligent Vehicles Symposium Proceedings, Dearborn, MI, USA, 8–11 June 2014; pp. 1037–1042. [CrossRef]
63. Merdrignac, P.; Shagdar, O.; Nashashibi, F. Fusion of Perception and V2P Communication Systems for the Safety of Vulnerable Road Users. *IEEE Trans. Intell. Transp. Syst.* **2017**, *18*, 1740–1751. [CrossRef]
64. Sewalkar, P.; Krug, S.; Seitz, J. Towards 802.11p-based vehicle-to-pedestrian communication for crash prevention systems. In Proceedings of the 2017 9th International Congress on Ultra Modern Telecommunications and Control Systems and Workshops (ICUMT), Munich, Germany, 6–8 November 2017; pp. 404–409. [CrossRef]
65. Sewalkar, P.; Seitz, J. Vehicle-to-Pedestrian Communication for Vulnerable Road Users: Survey, Design Considerations, and Challenges. *Sensors* **2019**, *19*, 358. [CrossRef] [PubMed]
66. Themann, P.; Kotte, J.; Raudszus, D.; Eckstein, L. Impact of positioning uncertainty of vulnerable road users on risk minimization in collision avoidance systems. In Proceedings of the 2015 IEEE Intelligent Vehicles Symposium (IV), Seoul, South Korea, 29 June–1 July 2015; pp. 1201–1206. [CrossRef]
67. Kaiwartya, O.; Cao, Y.; Lloret, J.; Kumar, S.; Aslam, N.; Kharel, R.; Abdullah, A.H.; Shah, R.R. Geometry-Based Localization for GPS Outage in Vehicular Cyber Physical Systems. *IEEE Trans. Veh. Technol.* **2018**, *67*, 3800–3812. [CrossRef]
68. Wymeersch, H.; Seco-Granados, G.; Destino, G.; Dardari, D.; Tufvesson, F. 5G mmWave Positioning for Vehicular Networks. *IEEE Wirel. Commun.* **2017**, *24*, 80–86. [CrossRef]
69. Bagheri, M.; Siekkinen, M.; Nurminen, J.K. Cellular-based vehicle to pedestrian (V2P) adaptive communication for collision avoidance. In Proceedings of the 2014 International Conference on Connected Vehicles and Expo (ICCVE), Vienna, Austria, 3–7 November 2014; pp. 450–456. [CrossRef]
70. Yahiaoui, S.; Omar, M.; Bouabdallah, A.; Natalizio, E.; Challal, Y. An energy efficient and QoS aware routing protocol for wireless sensor and actuator networks. *AEU Int. J. Electron. Commun.* **2018**, *83*, 193–203. [CrossRef]
71. Khairnar, V.; Kotecha, K. Performance of Vehicle-to-Vehicle Communication using IEEE 802.11p in Vehicular Ad-hoc Network Environment. *Int. J. Netw. Secur. Its Appl.* **2013**, *5*. [CrossRef]

© 2020 by the authors. Licensee MDPI, Basel, Switzerland. This article is an open access article distributed under the terms and conditions of the Creative Commons Attribution (CC BY) license (http://creativecommons.org/licenses/by/4.0/).

MDPI
St. Alban-Anlage 66
4052 Basel
Switzerland
Tel. +41 61 683 77 34
Fax +41 61 302 89 18
www.mdpi.com

Journal of Sensor and Actuator Networks Editorial Office
E-mail: jsan@mdpi.com
www.mdpi.com/journal/jsan

www.ingramcontent.com/pod-product-compliance
Lightning Source LLC
LaVergne TN
LVHW070559100526
838202LV00012B/512